Tropical Forestry

Tropical Forestry
Volumes Already Published in this Series

Harvesting Operations in the Tropics by Sessions, J. 2007
ISBN: 3-540-46390-9

Forest Road Operations in the Tropics by Sessions, J. 2007
ISBN: 3-540-46392-5

Tropical Forest Genetics by Finkeldey, R., Hattemer, H. 2007
ISBN: 3-540-37396-9

Sampling Methods, Remote Sensing and GIS Multiresource Forest Inventory by Köhl, M., Magnussen, S., Marchetti M. 2006
ISBN: 3-540-32571-9

Tropical Forest Ecology - The Basis for Conservation and Management by Montagnini, F., Jordan C. 2005
ISBN: 3-540-23797-6

John Sessions (Ed.)

Harvesting Operations in the Tropics

With 82 Figures and 19 Tables

 Springer

Volume Editor:

Dr. John Sessions
Department of Forest Engineering
Oregon State University
213 Peavy Hall
Corvallis
OR 97731-5706
USA

ISSN: 1614-9785
ISBN-10 3-540-46390-9 Springer-Verlag Berlin Heidelberg New York
ISBN-13 978-3-540-46390-0 Springer-Verlag Berlin Heidelberg New York

Library of Congress Control Number: 2006935981

Springer-Verlag is a part of Springer Science+Business Media

springer.com

© Springer-Verlag Berlin Heidelberg 2007

Editor: Dr. Dieter Czeschlik, Heidelberg
Desk Editor: Anette Lindqvist, Heidelberg
Production: SPi
Typesetting: SPi
Cover Design: Design & Production, Heidelberg

Printed on acid-free paper   31/3152-HM   5 4 3 2 1 0

Preface

Harvesting includes all the activities to fell trees and remove them from the forest to the roadside for loading and transport from the forest. Harvesting and extraction operations are the activities that generally cause the most significant impacts on forest managed for timber production. Sustainable forest management is concerned with management of forests in such a way as to control the impacts associated with harvesting and timber extraction. Harvesting and extraction for sustainable timber production in natural forests are not to be confused with logging associated with land conversion activities such as conversion to permanent or temporary agriculture, pasture land, or domesticated trees.

Much of the impact of harvesting and extracting can be reduced through proper planning and control of harvesting operations using principles, systems, and techniques common to temperate forests. However, many areas of the tropics pose unique operating conditions: heat, high humidity, high-intensity precipitation, occurrence of certain diseases, and lack of rock for suitable road surfacing. In natural forests, clear felling is rarely practiced, and many species are the rule with few commercial species on any hectare; thus, log removals per unit area are low.

Increasingly, planted forests are being established as a source of wood and fiber on previously cleared lands. These planted forests have high potential to contribute to wood supply, to reduce pressure on natural forests, and to contribute to economic development. Planted forests in the tropics pose special challenges. In many areas of the tropics, much unskilled or lower-skilled labor is unemployed, or underemployed. An important question for management of planted forests is the choice of appropriate technology to meet socioeconomic objectives.

The purpose of this book is bring together information on harvest methods, system productivity, and methods for conducting safe, efficient, and environmentally acceptable operations in tropical forests. It highlights the challenges of harvest operations in the tropics, includes techniques that have been shown to be successful, and discusses newer technologies. It is intended as a reference book for the forest engineer and others interested in planning and management of tropical forests. Numerical examples are included, where appropriate, to provide clarity for interpreting graphs, procedures, and formulas. The book is divided into ten chapters which cover the various facets of harvest systems from planning for felling and skidding to log transport from the forest, and concludes with considerations in selecting the appropriate harvesting technology.

Chapter 1 puts harvesting operations in context of forest management, defines the requirements for successful harvest operations, and reviews why operating conditions in the tropics are different from those for other forests.

Chapter 2 defines levels of planning, available mapping and information systems generally available in countries with tropical forests, and lays the important environmental and economic considerations for successful harvest operations in the tropics.

Chapter 3 covers organization, administration, and factors affecting worker safety and productivity in the tropics.

Chapter 4 reviews felling, bucking, and delimbing equipment, techniques, and production rates for a range of technologies from hand tools to highly mechanized specialized felling machines. Prefelling operations in natural forests and techniques for maximizing tree value in both natural and planted forests are addressed.

Chapter 5 reviews various animal extraction systems that have been used and continue to be used in areas where physical demands, social, and economic conditions make them appropriate.

Chapter 6 focuses on ground-based mechanized equipment for extraction, including rubber-tired skidders, flexible and rigid track skidders, rubber-tired forwarders, and their productivity. Skid trail patterns, landing design and operations, road spacing, cost, and environmental considerations are covered.

Chapter 7 focuses on systems suitable for steep land extraction. The various types of cable systems are defined with their planning requirements and factors that affect productivity. Considerations for helicopter operations are briefly reviewed.

Chapter 8 addresses a range of loading technologies from low-technology ramp systems through mechanized hydraulic swingboom and front-end loaders.

Chapter 9 reviews transport from the log landing to the manufacturing or shipping facility by truck, water, or rail. Common truck configurations, safe operating guidelines, and productivity are discussed. Water transport plays a larger role in the tropics than in other regions. Raft construction and use on rivers, swamps, lakes, and tidal forests are presented.

Chapter 10 discusses factors in selecting the appropriate harvesting technology, including local laws, customs, worker availability, tree size, and silvicultural system.

This book represents a compilation of available literature and the professional experiences of the authors. In particular we would like to recognize the long-term contributions of the Food and Agriculture Organization for promoting improved management of world forests, and their funding and documentation of many studies in tropical forest management and conservation.

September 2006   John Sessions

Contents

List of Authors

John Sessions
Department of Forest Engineering, Oregon State University, 223 Peavy Hall, Corvallis, OR 97331-5706, USA

Kevin Boston
Department of Forest Engineering, Oregon State University, 23 Peavy Hall, Corvallis, OR 97331-5706, USA

Glen Murphy
Department of Forest Engineering, Oregon State University, 271 Peavy Hall, Corvallis, OR 97331-5706, USA

Michael G. Wing
Department of Forest Engineering, Oregon State University, 275 Peavy Hall, Corvallis, OR 97331-5706, USA

Loren Kellogg
Department of Forest Engineering, Oregon State University, 261 Peavy Hall, Corvallis, OR 97331-5706, USA

Steve Pilkerton
Department of Forest Engineering, Oregon State University, 53 Peavy Hall, Corvallis, OR 97331-5706, USA

Johan C. Zweede
Tropical Forest Foundation, Rua dos Mundurucus 1613, Bairro, Jurunas, CEP 66025-660, Belém-Pará, Brazil

Rudolf Heinrich
Via Gorgia di Leontini 260, 00124 Rome, Italy

Introduction

1.1
Harvest Operations in Context

Forests cover 25% of the Earth's surface and 60% of the world's forests are within the tropics. More than one billion people depend on tropical forests for some part of their livelihood. Forests provide the opportunity to directly contribute to the well being of many and through economic development, and supporting industries, the well-being of many more. At the United Nations Conference on Environment and Development (UNCED) in 1992, the international community accepted the concept of sustainable development as a framework for future development. Sustainable development embodies a reduction of poverty, unemployment, and inequality.

Forest management is a deliberate and guided intervention to manipulate the complex biotic and human components of forests in order to satisfy the needs of particular beneficiaries. Forest operations consist of all technical and administrative processes to develop technical structures and facilities, to harvest timber, to prepare sites for regeneration, and to maintain and improve quality of stands and habitats. The concept of sustainability in forestry implies the total welfare effects of forest management should never decrease. Harvesting and extraction for sustainable timber production is not to be confused with logging associated with land conversion activities such as conversion to permanent or temporary agriculture, pasture land, or domesticated trees.

Harvesting is an essential activity in forest management. It involves all operations from tree felling to delivering logs at the mill, rail depot, or ship dock. If carefully planned and implemented, benefits, which were anticipated at the time of the forest investment, are possible. Poor planning and/or poor implementation of forest plans can be costly, result in environmental degradation as well as excessive wood waste, poor utilization of the available resource, and injury to personnel. Harvesting activities must consider, and can influence, stand treatment regimes, and must be consistent with terrain characteristics. Sound forest operations require a thorough understanding of these factors.

Historically, harvesting of tropical forests has been in natural forests, but this picture is rapidly changing. Plantations are a rapidly growing source of world wood, and the tropics are no exception. Today, forest plantations ring the globe, from Southeast Asia, to India, to Africa, and tropical Latin America. Most of these plantations are being grown for timber production.

This book presents the fundamentals of harvest operations both in natural forests and in planted forests. Our focus is on operations in wet tropical forests, but we provide information that can also be applied to dry tropical forests. The book is designed to be a handbook for the forest engineer and a textbook for those interesting in learning about tropical harvest operations. The fundamentals of road operations in natural and planted forests in the tropics are presented in a companion volume.

1.2
Overview of the Tropics

The tropics are bounded roughly to the north of the equator by the Tropic of Cancer and to the south of the equator by the Tropic of Capricorn. Although the tropics are often portrayed as an area that has remained unchanged for a long time, i.e., not subject to glacial activity and large climate swings, that does not mean that they are homogeneous. Differences can be described by elevation, topography, temperature, mean annual rainfall, seasonal rainfall, and soils. The tropics can be divided into the warm tropics and cold tropics, wet tropics and dry tropics. There are three major tropical regions, America, Asia, and Africa (Fig. 1.1) Some generalizations can be made between regions and within regions. Southeast Asia is much more mountainous than Africa or South America and has many young, eroding landscapes. It also has substantial areas of volcanic rock and soils. In Africa and South America, recent volcanic rock and soils are much less common.

Considering within-region variation, slightly more than 50% of tropical America, for example, has a mean annual precipitation of 1,600–3,200 mm/year and less than 5% receives more than 3,200 mm/year and less than 5% receives under 400 mm/year. Similarly, the majority of tropical America has pronounced seasonal rainfall, with most of the seasonal rainfall coming in summer, but some eastern coastal areas have the pronounced seasonal rainfall coming in winter. Other areas have rainfall well distributed throughout the year. One can generalize that tropical America is flat. Eighty-two percent is less than 8% slope, and 4% is greater than 30% slope. And one could generalize that tropical America is well drained, but about 25% of the flat terrain would be classified as poorly drained.

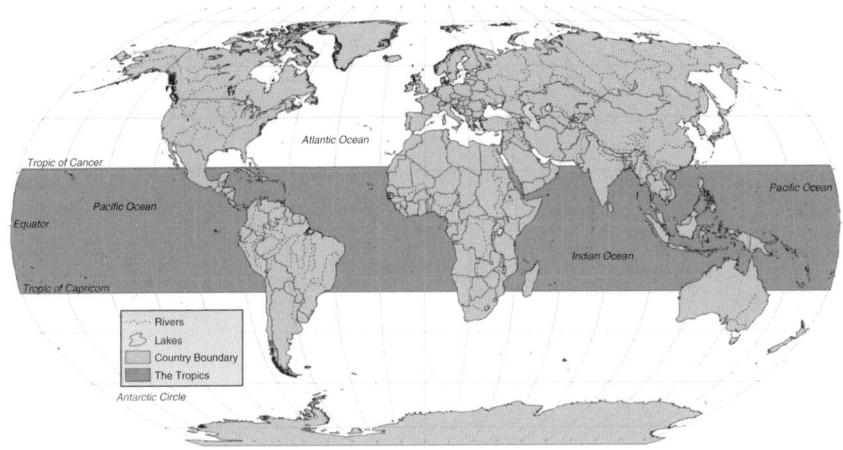

Fig. 1.1. Tropics of the world

Much international interest focuses on that part of the tropics that supports tropical rainforests (Fig. 1.2). Tropical America has about 50% of that total, with the remainder concentrated in Southeast Asia and Africa (Table 1.1) Tropical rainforests are found in places that have no dry months or only a few dry months. Tropical evergreen rainforest occurs in places with no dry season, and tropical semievergreen rainforest forms where there is a dry season. Tropical evergreen rainforests differ from semievergreen rainforests in that semievergreen rainforests have some species that are deciduous. In general, the western Amazon rainforests are evergreen, and the eastern Amazon rainforests are semievergreen. Southeast Asian rainforest is almost entirely evergreen and African rainforest is almost entirely semievergreen.

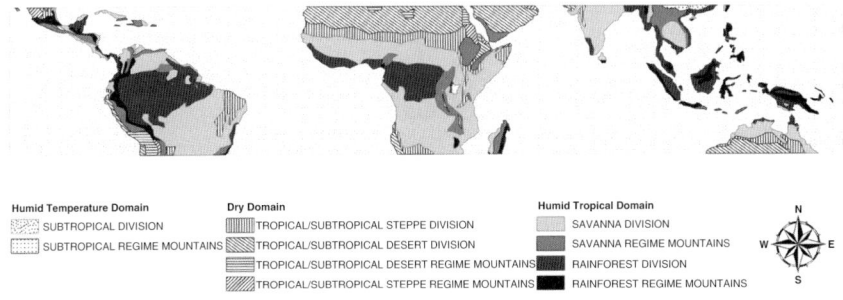

Fig. 1.2. Bailey's ecosystem classification of the tropics

Table 1.1. Distribution of tropical rainforest

Region	Rainforest area (million hectares)	Percentage of total	Primary location
South America	400	48	Amazon, Orinoco basins
Southeast Asia	250	30	Malesia, Asian subcontinent to India
Africa	180	22	Zaire basin to Gulf of Guinea

1.3
Factors Affecting Operations in Tropical Rainforests

There are some physical, biological, and social factors which, in combination, are unique to tropical forests:

- *Climate.* During much of the year the climate is often wet, hot, and humid, affecting choices of equipment, length of operating season, and labor productivity. Rapid wood deterioration after cutting may be a concern in some species.
- *Topography.* The topography is mostly flat, but mountainous regions exist in some areas. In many instances in the tropics, the terrain is hilly, consisting mainly of short hills. Water transport often plays a more important role for access and transport than in temperate areas.
- *Soils.* Wet soils provide poor traction for skidding and hauling. Sound rock is often not available, particularly in South America and Africa. Road surfacing may be expensive or scarce, affecting season of operation, road cost, and skidding distances.
- *Species.* In natural forest, up to 100 or more species can be found per hectare with widely differing tree sizes. Nearly all are hardwoods. Harvestable volumes of commercial species may be sparse, perhaps only several trees per hectare. An exception is the dipterocarp forest of Southeast Asia. Many of the trees have intermingled crowns connected by vines or lianas, and often supporting large hidden dead branches. At the base of many species are large flutes or buttresses.
- *Location.* Many tropical forests are in remote, undeveloped areas. Special attention must be given to choice of the level of mechanization and establishment of infrastructure for equipment maintenance, parts, and fuel supply.

- *Labor.* In some areas, a large unemployed or underemployed pool of unskilled labor may be available. Social policy as well as cost factors may affect the choice of harvesting systems.
- *Disease.* Tropical forests are home to parasites and require special planning to provide safe housing for forest workers. In particular, much of the tropics is affected by malaria, at least during certain periods of the year. In Africa, bilharzia eye disease from stagnant water is very frequent.

1.4
Successful Harvest Operations

Harvest operations to be successful must be:

- Technically feasible considering physical laws, engineering knowledge, and environmental relationship of the forest
- Economically viable considering the costs and benefit of short-range and long-range consequences
- Environmentally sound considering impacts on the natural and social environment, and efficient use of natural resources, including renewable materials, nonrenewable materials, water, energy, and space
- Institutionally feasible considering the laws and regulations governing operations, landowner objectives, and social values

1.5
Codes of Harvesting Practices

Considerable progress has been made in recent years in the introduction of environmentally sound forest harvesting practices in many parts of the tropics. A number of tropical countries have developed harvesting codes of practice. The International Labor Office (ILO) and Food and Agriculture Organization of the United Nations (FAO) have been active supporters.

One of the earliest national harvesting codes of practice in the tropics was adopted by Fiji in 1990 with the help of ILO. In 1996, FAO released its *Model Code of Harvesting Practice* (Dykstra 1996) which as of 2006 has been used as the template for at least 30 countries in the tropics. Several regional codes have been developed or are in development. Examples include the regional *Code of Practice for Forest Harvesting in Asia-Pacific* published by FAO in 1999 (FAO 1999) and endorsed by the Association of Southeast Asian Nations

(ASEAN) in 2001 and the *Regional Code of Practice for Reduced-Impact Forest Harvesting in Tropical Moist Forests of West And Central Africa* published in 2005 (FAO 2005).

The primary objectives of the harvesting codes are to promote forest harvesting practices that improve standards of utilization and reduce environmental impacts, thereby contributing to the conservation of forests through their wise use. Harvesting codes typically contain information on harvest planning, forest road engineering, cutting, extraction, landing operations, transport, postharvest assessment, and training, supervision, and safety for the forest work force.

Planning

Harvest planning in tropical forests shares many of the principles common to harvest planning anywhere. It involves a study and thorough understanding of:

- Objectives of the overall forest management plan
- Physical capabilities of the harvesting systems and the relationships between roads and harvesting systems
- The inventory of relevant site conditions, including climate, soils, vegetation, and topography

2.1
Levels of Planning

Forest planning can be divided into three levels: strategic planning, tactical planning, and operational planning.

Strategic planning involves high-level planning goals over broad areas and long periods of time. For public agencies, it may involve allocation of land to land use (zoning), establishment of concessions, and tests for sustainability. It may also involve establishment of standards and guidelines regarding management of riparian zones, wildlife corridors, and management of steep areas. For private companies, it may involve purchase of land, decisions to build facilities, and long-term decisions on timber supply, including the intensity of management.

Tactical planning involves shorter time frames and considers action at the landscape level, often the watershed level. Tactical planning draws its direction from strategic plans, but with the goal to develop spatially feasible plans that implement the direction from the strategic plan. Decisions at the tactical level involve development of management compartments, planning of primary road systems and main bridges, decisions concerning road density, and the preliminary scheduling of compartments to be entered based upon mill demands, market forecasts, and estimates of timber inventory.

Operational planning, sometimes called harvest planning, refers to planning the set of actions needed to implement specific projects on the ground. The time frame for operational planning is short, 2–3 years. Operational planning involves identification of specific compartments to be harvested, timber inventory of areas to be harvested, identification of environmentally sensitive areas that are not to be harvested, planning of secondary roads, landings, and skid trails. In compartments that are to be selectively harvested, trees are individually identified with paint, flagging, or tags. In compartments that are to be clear-felled, compartment boundaries are marked and specific trees or groups of trees to be retained are identified. In the year before harvest, secondary roads are constructed and crew scheduling plans are developed. Operational plans involve planning for both wet and dry seasons. Timing of road construction activities is particularly critical. Operational planning also involves crew scheduling to recognize equipment needs and productivity differences between wet and dry seasons, log inventory levels to reduce supply risks, mill demands, and contracts. At the production level, most firms implement quarterly, monthly, and weekly monitoring in order to adapt plans to changes in weather, crew productivity, mill demand, or markets.

2.2
Mapping Tools and Information Systems

Planning of forest operations requires the spatial location of vegetation types, soils, existing transportation networks, ridge lines, principal rivers, and hydrologic features, including waterfalls or rapids on important rivers. Mapping documents vary from one country to another, but generally include:

- 1:200,000 quadrangle maps with or without contour lines
- 1:200,000 thematic maps: hydrology, geology, soils, physiographic features, vegetation, political boundaries
- 1:50,000 quadrangle maps with contour lines (20-m intervals), special area maps of reserves or parks
- 1:50,000 Landsat or SPOT satellite imagery
- 1:200,000 radar imagery
- Historic and recent aerial photo coverage at a scale of 1:20,000 to 1:50,000

In spite of any imperfections, these maps can be a considerable help. Although some agencies may include accuracy assessments of the maps they produce, the actual number of control points checked is likely to be a small and reconnaissance will usually be necessary to confirm map feature locations.

Increasingly, hard-copy maps are often converted into a digital format through scanning processes and can be accessed and analyzed through a geographic information system (GIS). A GIS is core technology software for most forest planning systems and is designed to organize and analyze spatial data. A key ability of a GIS is the association of tabular or database information with geographic locations. GIS users are then able to select a location and view all tabular data associated with the selection, or reverse the selection process and see all locations associated with a database record selection. One of the strengths of a GIS is that it allows users to work with multiple files, often called layers, simultaneously, with each layer representing a separate feature such as roads, hydrology, or forest cover. Data analysis can then consider entire landscapes rather than individual components.

Most GIS use three geometric shapes to represent landscape features: points, lines, and polygons. Points are often used to locate features that occur at a single location and can be represented by a pair of coordinates. Examples of point features include measurement control points or survey monuments, wildlife nest sites, or culverts, depending on map detail. Lines are commonly used to describe linear features such as streams, roads, or utility features such as power cabling or pipe networks. Polygons are used to represent features that cover areas such as stand boundaries or soil types. Some GIS can accommodate imagery, also called raster data. Raster data refers to spatial data that are organized by cells, or pixels, rather than by discrete point, line, and polygon features. Maps produced from remote-sensing techniques such as from space- or aerial-based platforms are typically stored as raster data, although there are other data sources such as terrain models that are also stored as raster data.

The greatest determination of the value of a GIS is the quality of data that are input into the GIS. There are a variety of tools and processes to create spatial data for GIS input. Digital tools such as range finders, theodolites, and total stations are also used to create GIS data. Hand-held and survey-grade GPS receivers with digital data loggers are now widely used to georeference points during aerial and ground reconnaissance. The continued improvement of GPS capabilities now allows GIS data to be brought to the field to support operational design of the harvest when previously hard-copy maps and photo interpretation skills were the primary field data sources. Outside of tropical forests, digital terrain models (three-dimensional digital images of the terrain) are becoming available for many areas. Digital terrain models will undoubtedly become more available in tropical forests as lidar (light detection and ranging) technology continues to improve and drives costs down. Advances in color aerial photography now allow fine detail (submeter resolution) of forest areas to be captured and input into a GIS within several days. Regardless of data origins, successful GIS analysis relies on input data accuracy and precision. Poor quality data will lead to poor analysis results, regardless of the GIS software that is used.

2.3
Environmental Considerations

Harvesting operations affect water, soil, plant, wildlife, and recreational resources of the forest. Plant, wildlife, and recreational resources are affected through the choice of silvicultural prescription, harvesting system, and utilization standards specified in the forest plan. They are also highly affected by the conduct of the harvesting operation, particularly regarding residual stand damage. The silvicultural prescription also affects water and soil resources, but the choice of harvest method and the conduct of the harvesting operation can have a major impact on soil and water resources. Major impacts can include soil compaction, soil erosion, increased sedimentation, water temperature, and chemicals in streams. Many of these impacts can be reduced through proper planning and supervision of harvesting operations.

Harvesting operations also affect wildlife by introducing access to the forest and by establishing camps within the forest. Operators of concessions can directly influence wildlife conservation through attitudes, policies, and programs that they establish and enforce.

2.3.1
Residual Stand Damage

2.3.1.1
Natural Forests

Residual stand damage is often the leading environmental issue with harvesting in natural forests. Natural tropical forests are rarely clearcut and are usually managed under a polycyclic system with reentries in 20–40-year intervals (Fig. 2.1). Typical stand conditions for a mature tropical moist forest resemble a reverse J curve such as Table 2.1. As each year passes, some trees move from the diameter class they are in to a larger diameter class and some trees within the diameter class die. Typically the natural mortality rate is higher in lower-diameter classes and decreases as tree diameter increases. Harvesting in natural forest typically removes some of the larger-diameter trees, the number depending upon the percentage of commercial species, growth rates, and the reentry cycle.

Without careful planning and implementation, damage to the residual stand can be significant, resulting in damage to 60% or more of the residual trees. This can cause significant impacts on the future yield of the forest. Although removals of some noncommercial trees can accelerate development of the commercial remaining trees, control of residual tree damage is important.

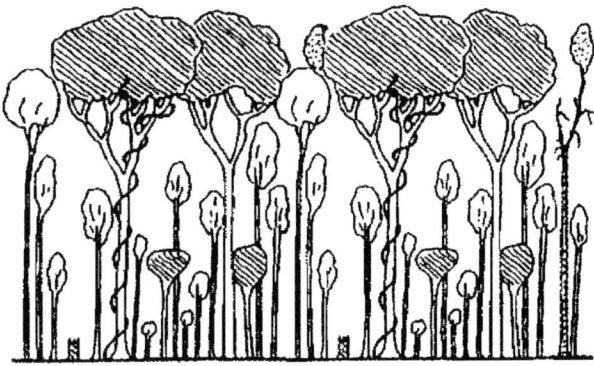

Fig. 2.1. Typical uneven-aged natural tropical forest with a few large trees and many smaller ones

Table 2.1. Pantropical stand table for primary moist forest. (After Dawkins 1959)

Diameter at breast height class (cm)	Number of Trees per hectare in class
10	242.0
20	97.0
30	40.0
40	19.0
50	11.0
60	6.8
70	4.5
80	3.3
90	2.3
100	1.5
>100	3.6

Felling of individual trees with large crowns can create gaps up to 0.20 ha per tree. Gaps exceeding 0.05 ha per felled tree are common. During felling, damage to regeneration near the base of the felled tree is often minor, but subsequent extraction can heavily damage regeneration. Without planning and control of extraction operations, skid trails over 80% of the harvested area have been observed. Damage to residual trees from skidding can often exceed damage from felling.

To limit residual stand damage, a number of forest practices have been developed and are sometimes referred to as part of "reduced impact logging" practices. Many of these practices evolved from temperate forest management, but some are unique to tropical forests. These practices include:

1. Conduct a preharvest inventory and map individual commercial trees.
2. Conduct preharvest planning of skid trails.
3. Cut vines 10–12 months prior to harvest in order for the vines to weaken enough to avoid danger and additional damage.
4. Construct skid trails to the minimum required size.
5. Use directional felling to minimize damage to future crop trees by felling the tree in a herringbone pattern to the skid trail using wedges or jacks as necessary.
6. Begin felling and skidding at the back of the harvest unit so that felled tree crowns are always behind the skidding operation and do not interfere with skidding.
7. Winch logs to planned skid trails and ensure that skidding machines remain on the skid trails at all times.
8. Use noncommercial species as bumper trees along the skid trail.
9. Fell any noncommercial trees for silvicultural improvement after the commercial trees have been felled and skidded.
10. With ground skidding, consider felling and skidding simultaneously (hot logging) to minimize obstructions to skidding.
11. With cable yarding systems, restrict yarding to skyline systems with slack-pulling carriages. Maintain at least partial suspension and full suspension where feasible. Do not use highlead systems in any polycyclic silvicultural system.

2.3.1.2
Planted Forests

As opposed to natural tropical forests, planted tropical forests are usually grown under monocyclic systems. Rotation ages are short, 40 years or less, with rotations less than 10 years being common for species managed for pulpwood. Residual stand damage is not an issue at final harvest, but during intermediate harvests can be an issue. Practices that have been found effective in reducing stand damage during intermediate harvests include:

1. Conduct preharvest planning of skid trails.
2. Construct skid and forwarding trails to the minimum required size.
3. When using skidders
   (a) Use directional felling to minimize damage to residual trees by felling the trees in a herringbone pattern to the skid trail.

    (b) Begin felling and skidding at the back of the harvest unit so that felled tree crowns are always behind the skidding operation and do not interfere with skidding.

    (c) Winch logs to planned skid trails and ensure that skidding machines remain on the skid trails at all times.

4. When using cable yarding systems

    (a) Restrict yarding to skyline systems with slackpulling carriages.

    (b) Maintain at least partial suspension, and full suspension where feasible.

    (c) Do not use highlead systems.

Tree spacing is a management issue with planted forests that will be harvested using mechanical harvesters and forwarders. If trees are planted less than 3.8 m apart, it is likely that the harvesters will not be able to pass between the trees without damaging the tree boles near the base, and if the trees are less than 3.5 m apart, an entire row will have to be removed. Options at tree planting time include planting at a rectangular spacing rather than a square spacing, for example, 3 m×4 m rather than 3.5 m×3.5 m, with the longer dimension in the direction the machine will pass.

    An open question is whether opening for skid trails reduces stand productivity. Evidence in temperate forests indicates that reductions in stand productivity are less than the percentage of the area in skid trails, owing to increased growth of trees along the trail. The narrower the skid trail, the lower the reduction in stand productivity.

2.3.2
Soil Compaction

Soil compaction results from pressures and vibrations applied to the soil. This increases the bulk density of the soil, reduces water infiltration, increases the penetration resistance that roots must overcome, and increases water runoff. Increased soil bulk density has been correlated with reduced tree growth in a number of studies. Soil compaction can be controlled through the choice of harvesting system, the season of operation, and postharvesting treatment or rehabilitation. If ground-based harvesting is chosen, soil compaction can also be limited by confining the area of land allowed to be in skid trails and landings. Logs or trees outside designated skid trails must be pulled to the trails by use of cables and the skidding equipment is not normally permitted to leave the trails.

2.3.3
Soil Rutting

Wheel and track ruts in forest soils are created when the soil cannot support the applied load from the wheels. These ruts and wheel tracks create pathways for water flow and subsequent erosion on sloping terrain as well as potential damage to tree roots. Ruts in skidder and forwarder trails are a function of soil strength, contact pressure, and number of trips. The use of soil strength measuring tools such as the cone penetrometer may allow location of the areas with the highest soil strength to reduce the likelihood of soil rutting.
Rutting can be reduced by:

- Restricting operations to periods when soil strength is greatest (lowest moisture)
- Locating trails above bogs and low places
- Increasing vehicle tire size, or track width, or use of wheel tracks
- Providing a mat of tops and smaller branches in the trail
- Controlling the number of trips along the trail through road and trail spacing
- Avoiding the use of chains
- Avoiding uphill skidding and high traction demands (large log loads) when soil strength is low

2.3.4
Soil Laterification

Concerns have been raised about soil hardening after clearing of roads, trails, landings, or after harvest through a process referred to as laterification. The largest soil groups of the tropics are Oxisols and Ultisols (referred to collectively as the "red" soils), which are older soils comprising 63% of the moist tropics, followed by Inceptisols (15%) and Entisols (14%), which are younger soils. The danger of lateritic formation after clearing is thought to be slight. In the Amazon, for example, it is estimated that about 6% of the region has the potential for laterification after clearing and that most of these soils lie in flat, poorly drained sites.

2.3.5
Soil Erosion

Erosion can result when soils are exposed to erosive forces such as water, wind, and gravity which create particle detachment and transport. Soil erosion can result from roads constructed for timber harvest, from landings, or skid trials. Soil erosion risks can be minimized by road location, road drainage, harvesting system choice, skid trail location, and postharvest activities, including water bars and seeding. Felling patterns, skidding direction, and season of operation are also important variables.

2.3.6
Stream Protection

Streams are an important resource for drinking water, irrigation, industrial use, fish, and a variety of wildlife. Harvesting activities along streams can affect water quality through sedimentation, increased temperature, and chemical contamination. One of the most effective methods for protecting streams is to designate special management practices along streams. Prior to harvest planning and preferably before stand establishment in planted forests, the forest management plan should designate areas where and under what conditions harvesting activities can occur. The treatment of unique areas adjacent to streams (riparian zones) depends upon a number of factors including:

- The reasons for protecting the stream
- The reasons for maintaining buffer strips
- The activities permitted in buffer strips
- The nature and importance of the indigenous cover
- The steepness of the slope
- The erodibility of the soil
- The intensity and amount of rainfall
- The width of the riparian zone

Harvest layout which favors uphill log movement away from streams is best for stream protection. During harvesting, riparian zones can be protected by controlling the felling direction through the felling method, avoiding skid road location near streams, by proper location of stream crossings, by controlling the season in which harvesting activities are done, and by postharvesting treatment in riparian zones, including removal of temporary stream crossings and removal of trees which inadvertently fell or were unavoidably felled into streams. Shade along streams is important in controlling stream temperature. Shade cover along streams can be controlled through harvest prescriptions in the riparian zone and zone width.

Crews should be instructed in spill prevention and cleanup procedures. Risk of introducing chemicals into streams from harvesting operations can be reduced by collection of engine oil and other equipment fluids during equipment maintenance. Equipment should be inspected for leaking seals and hoses. Equipment should not be washed in streams.

2.3.7
Wildlife Conservation

Harvesting can affect wildlife by (1) removing or degrading wildlife habitat and (2) by increasing access for hunting of wildlife. Habitat conservation can be affected by specification of reentry cycles for selective harvesting, limits on timber removal per entry, requirements for wildlife corridors, and protection of riparian areas. Habitat conservation policies are established by governments and are monitored for compliance through regional and local forest officials. Habitat conservation policies are relatively easy to monitor for compliance since trees are stationary.

Increasing markets for bushmeat and animal products have dramatically increased pressure on wildlife in many tropical forests. Governments also establish policies on hunting of wildlife for bushmeat and animal products. Hunting policies are more challenging to enforce. Concession operators can make important contributions to discouraging poaching and illegal trade in animal products through:

- Installing manned barriers at road entrances to the concession
- Blocking inactive forest roads with deep ditches or large logs
- Prohibiting outside lodgers in company camps
- Prohibiting setting up of camps in the concession area
- Reporting infractions to the authorities
- Prohibiting hunting by company employees outside of subsistence needs
- Prohibiting transport of noncompany persons or materials in company vehicles
- Making domesticated animal protein available at reduced prices at company camps
- Making severe penalties for violations of company rules
- Raising local awareness of the need to use wildlife resources sustainably

2.4
Wood Utilization

Good wood utilization has an important impact on harvesting cost and the overall profit margin. Product manufacture begins with the felling operation. Lowered stump height, reducing felling and skidding breakage, and correctly bucking trees into logs meeting buyer or mill specifications all increase the yield from the forest. Increased yield per hectare reduces the fixed cost per unit of wood delivered. More importantly, increased yield increases the gross revenue and overall profitability of the operation.

2.5
Costs of Environmental Protection

Proper planning and control not only improves environmental performance in both natural and planted forests, but also reduces costs.

2.5.1
Natural Forests

In natural forests proper planning and control involves a set of activities recently referred to as reduced impact logging (RIL). These include:

- Preharvest inventory and mapping of trees
- Preharvest planning of roads and skid trails
- Preharvest vine cutting
- Directional felling
- Cutting stumps low to the ground
- Efficient utilization of felled trees
- Constructing roads and skid trails to optimum width
- Winching of logs to planned skid trails
- Constructing landings of optimal size
- Minimizing ground disturbance and slash management

If proper planning and control is not implemented, then harvesting is done on a "hit or miss" approach resulting in one or more of the following outcomes:

- Potentially missing valuable trees
- Cutting of nonmerchantable trees
- Excessive ground disturbance
- Excessive damage to the residual stand
- Poor utilization of trees that are cut
- Loss of merchantable logs because crews cannot find them
- Poor utilization of skidding machines
- Damage to skidding machines
- Higher logging costs
- Lower net revenues

The benefits of proper planning and control in natural forest operations were documented in a study in the eastern Amazon (Holmes et al. 2000). There, an initial harvest of about 25 m³/ha (four to six trees) was removed in controlled areas where (1) proper planning and control was done and (2) little planning and control was done. With proper planning and control:

- More careful tree selection resulted in a decrease of 1.4 m³/ha of logs that were cut but were not merchantable.
- More careful bucking of logs recovered an additional 1.1 m³/ha.
- More careful skidding of logs recovered an additional 0.9 m³/ha.
- There was a 56% reduction in the number of residual commercial trees fatally damaged.
- There was a 50% reduction in the area disturbed by heavy equipment.
- There was a 12% reduction in overall cost per cubic meter.

Proper planning and control requires higher "upfront" costs for inventory, mapping, and planning of skid trails (Table 2.2). Felling and bucking costs are also higher for directional felling and improved recovery. However, there are large efficiencies in skidding and log deck operations. Stumpage costs were decreased through increased recovery as stumpage costs were levied on a per hectare basis.

Table 2.2. Cost and revenue summary from an eastern Amazon study. (Holmes et al. 2000)

Activity	Average cost without planning and control (US$/m³)	Average cost with planning and control (US$/m³)
Preharvest Planning		
Block layout	–	0.26
Inventory	–	0.48
Vine cutting	–	0.14
Data processing	–	0.10
Mapmaking	–	0.20
Harvest Planning		
Tree hunting	0.14	–
Tree marking	–	0.13
Road planning	–	0.02
Log deck planning	–	0.01
Skid trail layout	–	0.27
Infrastructure		
Road construction	0.28	0.16
Log deck construction	0.29	0.16
Logging Felling and bucking	0.49	0.62
Skidding	1.99	1.24
Log deck operations	2.01	1.28
Waste adjustment	0.40	0.09
Stumpage cost	9.09	7.61
Training	–	0.21
Overhead/support	0.97	0.86
Total costs	15.66	13.84

2.5.2
Planted Forests

With planted forests, the major link between harvesting and environmental protection is that wood from planted forests reduces the pressure on natural forests through substitution. Productivity from planted forests is typically much higher than from natural forests, so production can be concentrated on a much smaller area. Issues regarding soil disturbance and water quality are similar to those for natural forests. Forests planted for timber production must be economically viable as they require upfront investments and products from planted forests compete in a global market. Proposed projects need to be carefully benchmarked against operations in other regions to verify that they are economically viable. Harvesting and transport are the largest costs for most planted forests; therefore, great care must be taken to identify suitable areas for planted forests and to conduct forest operations in an efficient way. Planted forests usually cannot directly compete against prime agricultural land; degraded agricultural or pasture lands are often good candidates. Planted forests are usually monocyclic although intermediate harvests for forests grown on saw timber rotations may be done.

Planning and control issues for planted forests include:

- Site productivity
- Tree size at intermediate and final harvest times
- Terrain (flat, steep, rolling, broken)
- Precipitation pattern
- Distance to mill or market
- Road construction costs, including availability of adequate surfacing materials
- Road, rail, or water transport infrastructure
- Appropriate harvesting technology considering terrain, tree size, local costs, social and political factors
- Road and landing spacing

2.6
Costs for Comparing Alternatives

2.6.1
Production and Cost

In order to develop an economical plan it is necessary to evaluate the cost of harvesting alternatives. This requires estimates of (1) harvesting production rates and (2) equipment and labor costs. To do this, the harvest planner must

Table 2.3 Machine-operating cost estimate. (After FAO 1977)

Machine:	Description _____
	Gross HP _____ Delivered cost _____
	Life in years _____ Hours (days): per year _____ life _____
Fuel:	Type _____ Price per litre _____
Tires:	Size _____ Type _____ Number _____
	Cost of replacement set _____
Operator:	Rate per hour (day) _____ Fringe benefits _____%

Cost Component			*Cost per hour (day)*
a) Depreciation	=	$\dfrac{\text{delivered cost} \times 0.90}{\text{life in hours}}$	_____
b) Interest	=	$\dfrac{\text{delivered cost} \times 0.60 \times \text{interest rate}}{\text{average hours per year}}$	_____
c) Insurance	=	$\dfrac{\text{delivered cost} \times 0.60 \times 0.03}{\text{average hours per year}}$	_____
d) Taxes	=	$\dfrac{\text{annual tax amount}}{\text{average hours per year}}$	
e) Operating Labor	=	$\dfrac{\text{labor cost per period} \times (1+f)}{\text{machine hours in period}}$	_____

where f = cost of labor fringe benefits expressed as % of direct labor cost.

		Sub-total[a]	_____
f) Fuel	=	GHP × X × CL	_____

where GHP = gross engine horsepower;
    CL = fuel cost per litre in dollars,
    X = 0.12 for diesel fuel, 0.175 for gasoline

g) Oils and greases	=	$\dfrac{\text{GHP} \times \text{X} \times 3.4^c}{100}$	_____

where X = 0.20 for tractors, skidders, front end loaders and trucks;
    X = 0.30 for feller-bunchers and knuckle boom loaders;
    X = 0.50 for processors, harvesters and forwarders.

h) Servicing and repairs[c]	=	$\dfrac{\text{delivered cost}}{\text{life in hours}^d}$	_____
i) Tires for hauling rigs	=	0.0006 × CST	_____

where CST = cost of set of replacement tires.

		Total[b]	_____

[a]This represents the *cost per standing hour* of a log hauling truck.
[b]This represents the *cost per traveling hour* of a log hauling truck, and cost per productive machine or effective hour for other machines.
[c]Include tires except for log hauling trucks.
[d]Use lifetime traveling hours in case of log hauling trucks.
[e]Cost multiplication update to 2006.

understand the operating and cost characteristics of the available logging systems especially with respect to tree size, harvest volume, skidding distance, and skidding direction. The planner must also understand the cost, production, and environmental interactions between road standards, road spacing, and skidding distance. Relationships for developing production rates are discussed in later sections. Equipment and labor costs are discussed next.

2.6.2
Costing Equipment and Labor

The hourly cost of the harvesting equipment and labor required to do a job is called the machine rate. The objective in developing a machine rate should be to arrive at a figure that, as nearly as possible, represents the cost of the work done under the operating conditions encountered and the accounting system in force. Most manufacturers of equipment supply data for the cost of owning and operating their equipment that will serve as the basis of machine rates. Caution should used though as these guides often assume the equipment is able to work thoughout the year. Therefore, such data usually need modification to meet specific conditions of operation, and many owners of equipment will prefer to prepare their own rates based on their own costs and production rates. The machine operating cost estimate form in Table 2.3 can be used as a guide in developing machine rates. This simplified guide assumes a salvage value equal to 10% of the delivered equipment cost and an average annual investment equal to 60% of the delivered cost. It is suitable for use with many mechanized operations. A similar form can be developed for operations using animals.

Organization, Administration, and Labor Productivity

Two of the main reasons for variations in efficiency between different enterprises are the quality of organization and the supervision and training of the work force. Proper planning and implementation requires organization and administration similar to any business. Safe and efficient production requires a well-trained, motivated work force.

3.1
Organization and Staffing

Each operation will require a staff for planning, organization, and instruction. The number of planners, administrators, and supervisors depends on the size of the operation. Seven functions are necessary:

1. Management to set goals, coordinate the work, and ensure the work is performed according to instructions and policy
2. Administration to be responsible for administrative work, including a wide field of aspects from housing, personnel, and economic transactions to clerical work
3. Planning to allocate resources in an efficient and environmentally acceptable fashion and to advise field personnel (and the general manager) on working standards
4. Operations to manage felling, skidding (yarding), and transport in a direct line position and to be responsible for the fulfillment of goals set by the general manager
5. Maintenance to be responsible for the efficient operation of the workshop and to provide field operations with quick repair and maintenance groups and keep on hand an adequate stock of spare parts
6. Supervision to supervise felling crews, tractors and trucks, and group work in various places
7. Training to train supervisors and workers how to safely and efficiently perform the tasks

These seven jobs are necessary in every operation. In very small operations all seven may be held by one person, whereas in larger operations each function may include several persons. The number included in each function may be derived from the following list, which is based on experience in well-planned and organized operations:

- *General manager.* There should always be a general manager, whatever the size of the operation.
- *Administrative manager.* The administrative manager is assisted by clerks and typists. If the total number of workers is less than 20, the general manager will normally also be the administrative manager. If the number of workers is 20–60, the administrative group is two or three persons. If the number of workers is 60–100, the administrative group is enlarged to five or six persons. If the workers are more than 100, the group increases by one person per 50 persons.
- *General planner.* With fewer than two managers for felling, skidding, and trucking, the general manager will also do the planning. If there are two or three felling, skidding, and trucking managers, a general planner is employed, sometimes with an assistant.
- *Managers for felling, skidding, and trucking.* There should never be more than one for each phase, i.e., not more than three in all. If there is only one group leader, he or she will also be the manager. If there are two group leaders, it might be necessary to have a separate manager. If there are three group leaders, there should be at least one manager. If there are more than four group leaders, there should be three managers.
- *Group leader-supervisor.* A good average is one group leader per ten to 20 workers. Where safety risks are present, one group leader per seven to ten workers is appropriate.

Examples

Operation 1: 300 workers	Operation 2: 50 workers
1 general manager	1 general manager
1 general planner	0 (or 1) general planner
10 administrators – clerks, etc.	3 administrators – clerks, etc.
3 managers	1 or 2 managers
20 group leaders	3 or 4 group leaders

3.2
Labor Training and Motivation

Timber harvesting in the tropics is largely dependent on the abilities, skills, and motivation of workers. Some operations are primarily manual work, while others use machines – the machines in turn are dependent on the operators.

Healthy, safe, and productive workers are necessary for efficient operations. The following approaches can promote a productive work force:

- Convince top management/owners that labor force improvements in safety, health, training, selection, and motivation yield returns in both the short term and the long term.
- Remove the barriers to human performance: poor water, heat stress, poor nutrition, poor tools and equipment, disease, and other obstacles that can be readily identified.
- Provide personal with protective equipment (suited to the tropics) and enforce its proper use.
- Analyze all jobs for safety hazards and minimize them through modifying the harvesting system and the machines along with the work practices.
- Match the jobs to worker capabilities and select workers capable of performing the tasks.
- Train supervisors and workers how to safely and efficiently perform the tasks.
- Make safety and health a primary management tool of productivity improvement. Consider the maintenance of the work force as important as the maintenance of the machines; consider labor force improvement as important as new machines.
- Align management, supervisor, and worker goals so they are mutually beneficial and supportive.
- Incorporate the findings of labor force and ergonomic research into the working conditions in the tropics.
- As a manager or supervisor, treat forest workers as you would like to be treated if you were a worker in their place.

Labor force improvement efforts can be assessed in a fashion similar to other improvement efforts in timber harvesting. For reasonable expenditures, labor force improvements return proportionally high returns compared with other expenditures. Safety and health experts, ergonomic specialists, and training institutions can provide assistance in designing and implementing labor force improvement programs.

Studies in temperate countries have found that forest equipment operators show a strong increase in productivity for the first 1.5 years and a slow but steady increase during the following 4 years. Turnover is generally considered to be one of the biggest obstacles toward achieving and maintaining a well-trained work force. A high level of turnover may discourage forest enterprises from investing in training. As the unit cost of production is directly a function of the worker productivity, effectively training a worker and retaining a trained worker is of high importance.

3.3
Worker Safety

Forest work is one of the most strenuous and dangerous occupations. It often involves heavy and demanding physical effort under hot, humid conditions along with exposure to insects and disease. Workers around equipment are subject to noise, vibrations, and equipment exhausts. Because of this, those planning, supervising, and doing forest work should know how to avoid unnecessary health risks and strain while working effectively and safely. Ergonomic requirements should enter the planning, organization, and supervision of forest work and ergonomic principles should guide the design, purchase, and maintenance of forest tools, equipment, and infrastructure. Due to remote work locations, all forest workers and their supervisors should receive intensive first aid and injury training by qualified instructors.

3.3.1
Energy Expenditure

Owing to the heavy nature of forest work and the relatively small body size of forest workers in many tropical forest areas, the expenditure of human energy must be conserved. This can be done through:

- Careful planning of the work to minimize unnecessary walking, handling, and work movements
- Arrangement of heavy work with light work in alternating stages
- Avoidance of static muscle strain by performing work with suitable muscle groups
- Maintenance of suitable work pace to keep the degree of strain between 40 and 50% of maximum working capacity
- Scheduling of rest breaks in relation to the heaviness of the work load, and ensuring ample resting time during these breaks
- Assignment of an adequate number of persons to do the work
- Providing suitable work tools

3.3.2
Nutrition

An average worker with a body weight of 55 kg, doing heavy forest work in a tropical climate, needs approximately 4,000 kcal of energy per day (Table 3.1). About 2,000 kcal can be spent on daily work and the rest to maintain biological functions. If food intake falls to 3,000 kcal, work performance will fall to about

Table 3.1. Example of daily food allowance for a worker weighing 55 kg doing heavy forest work in a tropical climate

Food type	Daily food allowance
Rice	600 g
Other cereals	200 g
Fish, meat	250 g
Eggs	2
Fats	50 g
Sugar	50 g
Fruit	200 g
Vegetables	200 g

50% of full performance. If caloric demand is not satisfied, working capacity will be reduced in the form of slow work movements, frequent pauses, and short working hours. Insufficient food intake may lower resistance to disease, and cause frequent accidents. Also, nutrition may influence absenteeism and labor turnover.

The food should contain salt. In hot climates water and other liquids are needed in quantities of 3–4 L/day. In the tropics this should be 5 L or more. Dehydration and its milder form hypohydration have both short-term and long-term health effects. In the short-term poor body hydration impairs cognitive performance, physical strength, and aerobic power, rendering the worker prone to injury and heat illness. In the long term the potential consequences of hypohydration are kidney stones and bladder cancer.

In high temperatures sweat evaporation is an important component of body cooling. The rate of sweating increases directly with the ambient temperature. The maximum sweat rate is 1.5–2 L/h. The maximum absorption of water is about 1.5 L/h, so any sweat loss beyond this rate may result in gradual dehydration regardless of the quantity drunk. In climatic conditions where the relative humidity is high, such as tropical forest, the sweat drips from the skin. This is caused by the inability of the already nearly saturated air to take up additional moisture from the skin. This loss of fluid from the body serves no cooling effect on the body but can severely deplete the body's fluids.

3.3.3
Heat Stress and High Altitudes

A forest worker generates body heat in proportion to energy expenditure. Heat stress results when the body cannot eliminate excess heat at the skin's surface. The usual circumstances for heat stress are high air temperature and high humidity, combined with low air velocity. These conditions restrict sweat evaporation, increase body temperature and pulse rate, and reduce working capacity.

Heavy work can be fully performed at temperatures of 25°C and 100% relative humidity. Above that temperature, working capacity is decreased. In tropical lowland forests, it is common to observe temperatures of 25–30°C, and humidity near to 100% (Table 3.2).

Measures to reduce heat stress include:

• Organizing work according to the changing heat of the day
• Scheduling sufficient rest breaks
• Providing appropriate clothing

High altitudes also affect worker productivity. For workers acclimatized to higher altitudes, capacity begins to decrease at elevations above 1,200 m. Working capacity is reduced about 8% at 2,300 m and 15% at 4,000 m. For workers not acclimatized to working in higher elevations the reduction in capacity is much greater.

3.3.4
Vibration

Vibration presents hazards for all forest workers, but particularly chainsaw operators. A great deal of resistance is produced when cutting through tough wood with a saw powered by a gasoline engine, and tropical hardwoods are among the toughest woods. Vibration leads to discomfort and fatigue. Few machines are more dangerous than a running chainsaw in the hands of a fatigued operator. In addition to fatigue, chainsaw operators exposed to saw vibrations over long periods of time run the additional risk of slow deterioration of circulation and destructive skeletal damage in the hands and arms. Vibration illness can modify the cardiovascular system, cause cystic changes in wrist bones, pain in the joints of the wrist, elbow, and shoulder, periodic headaches, and general irritability. Modern chainsaws include antivibration systems and proper saw maintenance and operating techniques can further reduce vibrations.

Vibration is a source of injury for other operators as well, including skidder, forwarder, and truck drivers. Back pain is particularly common. With skidders

Table 3.2. Estimated reduction in productivity levels due to heat stress

Temperature (°C)	Reduction in productivity (%)
26	0
28	10
30	20
32	35
34	65

and forwarders the only shock absorbers are the tires and the operator's seat. Selection of a suitable seat can be very important for worker safety and productivity. Trucks have shock absorbers, but their higher speed also makes vibration an important safety and productivity consideration. For trucks, maintaining road smoothness and operating at reduced tire inflation pressures can substantially reduce vibrations to both vehicle and operator.

3.3.5
Noise

Machines powered by internal combustion engines generate noise both by their mechanical operation and by the exhaust system. High noise levels create fatigue, irritability, and hearing damage and prevent operators from hearing warning shouts from other workers. Hearing protection should be used. Breaking up tasks also helps, so that the cutting time is not continuous. Supervisors should verify that saws are never run without mufflers, both for worker safety and for fire protection.

3.3.6
Exhaust Gases

Chainsaws use two-stroke gasoline engines. Two-stroke engines produce large amounts of carbon monoxide, compared with four-stroke engines. As much as 30% of the fuel flowing through them is emitted unburned or partially burned into the air in the form of carbon monoxide and other undesirable chemicals. When inhaled, carbon monoxide blocks the blood's ability to carry oxygen to vital organs. Depending on the extent of the exposure, symptoms can range from a mild headache to nausea, dizziness, intense headache, unconsciousness, and death. In order to reduce the risks of chronic exposure to carbon monoxide, supervisors and workers should ensure the saw they use is in good condition, that it contains the correct oil-to-gas mixture as recommended by the manufacturer, that the engine is well tuned, and that the muffler is working well.

Most heavy forestry machinery uses diesel fuel in four-stroke engines. When diesel fuel burns in an engine, the resulting exhaust is made up of soot and gases representing many different chemical substances. Most of the soot consists of particles less than 2 microns in diameter that can be inhaled and deposited in the lungs. Diesel exhaust contains many more particles than gasoline exhaust and includes potentially cancer causing substances. Gases in diesel exhaust can also create health problems. Eight of these gases are nitrous oxide, nitrogen dioxide, formaldehyde, benzene, sulfur dioxide, hydrogen sulfide, carbon dioxide, and carbon monoxide. Enclosed cabs with filtered air or with air conditioning provide the best operator protection. Operators can wear respirators to filter out oily particulates and organic compounds, but they are less satisfactory. Equipment maintenance shops should be well ventilated.

3.3.7
Operator Safety Equipment

Operators should use appropriate safety equipment. Depending on the specific task these should include a hard hat, safety boots, safety goggles, gloves, trim fitting clothes, leg protective devices, and a high visibility vest or high visibility clothing.

3.3.8
Insect Control

Although workers are exposed to insects bites in all forest areas, the tropics have several diseases that require particular care: malaria, leishmaniosis, African sleeping sickness, and yellow fever.

Malaria is endemic to large parts of tropical forests. Workers need to be protected from mosquitoes that transmit malaria. Protection includes:

- Care in locating worker camps away from standing water
- Eliminating places around camps where mosquitoes breed
- Spraying dwellings with insecticides
- Sleeping under bed nets – particularly effective if they have been treated with insecticide
- Wearing insect repellent and long-sleeved clothing if out of doors at night

In some parts of the tropics leishmaniosis is endemic. Workers need to be protected from the sandflies that transmit it. Reservoirs of the disease include rodents and dogs. Protection includes:

- Spraying dwellings with insecticides
- Controlling rodent populations
- Eliminating rubbish heaps which are sandfly breeding areas
- Using insect repellents and protective clothing
- Keeping dogs indoors after sunset and removing infected dogs

In parts of tropical Africa the tsetse fly spreads African sleeping sickness (trypanosomiasis). The initial signs are initially nonspecific (fever, skin lesions, rash, edema, or lymphadenopathy); however, the infection progresses to meningoencephalitis and untreated cases are eventually fatal. No vaccine is available to prevent this disease. Tsetse flies are attracted to moving vehicles and dark, contrasting colors. They are not affected by insect repellents and can bite through lightweight clothing. Areas of heavy infestation tend to be sporadically distributed and are usually well known to local residents. Protection includes:

- Avoiding known areas of infestation
- Using clothing of wrist and ankle length made of medium-weight fabric in neutral colors that blend with the background environment
- Monitoring worker condition and providing access to medical treatment

In tropical forests, mosquitoes also spread yellow fever. Vaccination is the most important measure. The fatality rate for infected unvaccinated workers is greater than 50%. National vaccination programs are under way in many countries, but there are still unprotected segments of the population in high-risk areas. Protection includes:

- Vaccinating all workers in high risk areas
- Monitoring workers for yellow fever outbreaks

3.4
Camp Facilities

Many harvesting operations are in remote areas and camps will need to be established. Camps need to be carefully planned as they house many different structures: administrative offices, workshops, equipment maintenance facilities, commercial buildings such as a market, social services such as a hospital, primary school, church(s), housing for all levels of workers, and a recreational area such as a soccer field. They must provide for fuel storage and dispensing stations, electric power distribution, water treatment and distribution, sewage treatment, waste sites, and communication facilities.

The camp can either be centralized or have a base camp center with satellite camps. Some camps may be located as part of an industrial development such as a sawmill or pulpmill, or may be deliberately located away from the industrial facility to higher elevations to provide access to cooling winds or to avoid areas where disease-carrying insects may be present during evening hours. Where necessary; insect spraying programs should be considered.

Worker living quarters should have clean running water, lighting and a power socket, and access to a shower and toilet draining to a septic tank. Local customs should be recognized such as the use of beds or hammocks. Refuse bins with animal-proof covers should be provided and refuse should be collected at least twice per week. To the extent possible, living quarters in permanent facilities should not be constructed of flammable materials.

Logging camps should be supplied with clean water, water should be tested regularly and treated if necessary, and upstream access should be protected.

Refuse sites should be carefully located away from water supplies and where winds will not blow smoke from burning refuse or carry odors to the camp. Access to refuse sites should be fenced if possible to prevent animals and children from entering. Household refuse should be burned periodically. Waste from the hospital should be handled separately. Septic tank effluent should be buried.

Camp roads should be lit at night and equipped with drinking-water points. Dust abatement should be considered along main camp streets during the dry season. Vehicle speed in camp should be controlled and use of truck engine brakes prohibited. Control of noise consistent with cultural norms should be enforced to provide for a civil society.

Felling

4.1
Introduction

In the past, felling in natural and planted forests was done with axes and cross-cut saws. In natural forests, axes and crosscut saws have largely been replaced with power saws. In planted forests, crosscut saws are still sometimes used for smaller timber and power saws are used for larger trees. Axes, machetes, and power saws are used for delimbing. In a few planted forests, mechanized felling has replaced the power saw following developments in temperate forests.

4.2
Field Planning

Felling is the first step in the conversion process of trees to forest products. The objectives of the felling and bucking process must be understood by supervisors, formen, and workers. Log specifications must be clearly specified and quality control maintained to minimize wood waste in the field and to improve recovery at the mill. Foremen and workers must understand how felling and bucking affects the efficiency of the skidding operation and profits to the mill.

Natural forests in tropical regions present special challenges. In natural forests, trees suitable for utilization are irregularly distributed and the volume and number per hectare are often low. Owing to the large number of species, their identification may be difficult. Each felling team should have an operation map (Fig. 4.1) which shows the approximate location of the trees to be felled, roads, landings, and skid trails. The map scale should be about 1:5,000. Where possible, skid trails should be flagged or identified before felling to facilitate directional felling. Directional felling should consider priority criteria such as future harvest trees, skidding direction and reduction of crown gap size.

Trees to be removed should have been painted, flagged, or tagged during preharvest planning. A multiple tag system can be used for identification, payment, and control. During the preharvest inventory, the inventory crew

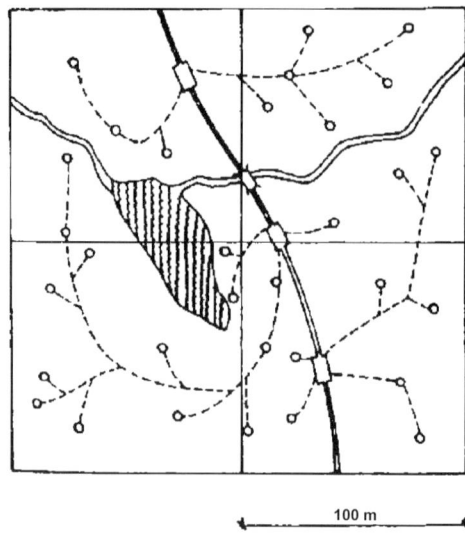

100 m

Fig. 4.1. Operation map for a felling team

nails three tags on each tree giving its tree number, compartment number, tree quality, and species. The cutter retains the first tag, the choker setter with the skidding operation retains the second tag, and a worker on the landing removes the third tag and paints log ends. The tags can be used for inventory control, chain of custody, and worker payment.

Trees that require special protection during felling and extraction for crop, cultural, wildlife, or other purposes should have been specially flagged during the preharvest inventory.

Foremen must carefully control the felling team working locations. Manual tree felling is dangerous anywhere, but in dense tropical forest it is especially dangerous. Tree crowns are large, often hiding dead branches, and dense undergrowth makes it difficult to retreat. Trees may be interconnected through climbers and tree centers may be hollow or rotten. When trees fall they frequently pull down other trees. Branches from falling trees and climbers may be broken off and snap backward. Felling teams must keep a minimum distance of two tree lengths from other workers. This distance should be increased to four tree lengths when visibility is poor. Two escape routes should be planned and prepared at a 45 degree angle away from the tree fall.

Under some conditions, the felling and skidding may need to be closely coordinated so that logs are removed before additional trees are felled. Large numbers of trees lying on the ground may impede extraction. They can also lead to lost logs, unnecessary breakage as trees are felled onto other trees, and unnecessary damage to the residual stand.

4.3
Manual Felling

4.3.1
Equipment

4.3.1.1
Axe

The axe has two basic functions, to cut and to split. The sharp edge of the axe makes the cut and the wedge-shaped head follows and opens the cut. There are different designs of axes depending upon the use for which the axe is intended (Table 4.1). Felling axes, used to chop into tree trunks, require a profile that both cuts and wedges. An axe used predominantly for limbing does not require the same tapered profile as the thickened taper which results in a splitting action. A splitting axe is normally heavier and wider than a chopping axe. Axes designed specifically for splitting often have a ridge on each side of the head which permits easy removal of the axe head from splits in logs or bolts.

There is a considerable variation in axe handles, from straight round handles for tools used in working in two directions, such as a splitting axe or a double-bitted axe, to oval curved handles recommended for felling and branching axes. An axe handle for chopping and branching axe weighs about 1–2 kg. Its length is between 70 and 80 cm and the length of the handle in relation to the user should equal the distance from the cutter's armpit to the finger tips, or from the hand to the ground when placed at the side while standing erect.

4.3.1.2
Crosscut Saws

Saws are used to fell trees and to cut them into bolts or log lengths. Saws waste much less wood than axe felling or bucking. Two-man crosscut saws are used for felling and bucking trees more than 30 cm in diameter. These saws have a long blade with two handles (Fig. 4.2). The standard length is 1.5 m and the usual blade thickness is 1.6–1.8 mm. The tooth line may be curved or straight

Table 4.1. Optimum weight of axe head

Description	Weight (kg)
For small trees and brush	0.7–0.8
For large softwoods	0.9–1.2
For large hardwoods	1.3–1.7
For large tropical woods	1.3–2.3

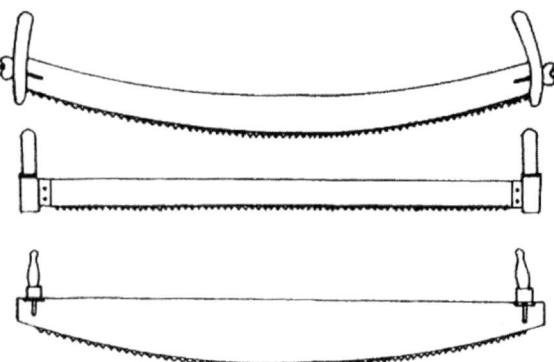

Fig. 4.2. Crosscut saws

and the back line may be curved or straight. A curved back line reduces saw weight and facilitates wedging. Sometimes taper grinding is used to make the back of the saw blade narrower to reduce saw friction. Teeth designs are of two types, peg or raker. Peg teeth are composed of only cutter teeth, while raker designs have cutters and rakers to clean the cut. Raker designs are more efficient if saw maintenance is adequate. Maintenance of raker-toothed saws demands skill, accuracy, and special tools which may not be available, making peg-toothed saws the only practical alternative.

As a general rule in handsaw operations, a team of two should work together. This involves the least amount of walking and delay and also reduces the accident risk. The following equipment is carried by the team:

1. First man
   (a) The crosscut saw
   (b) A protective cover attached to the saw when walking from one tree to another
   (c) A machete with a protective cover
   (d) An axe (weight about 1.5 kg.; length of handle about 70–80 cm with a protective cover)
   (e) A hard helmet in a bright color
   (f) A first-aid pocket kit
   (g) A whistle
2. Second man
   (a) A sledge
   (b) Three wedges
   (c) A machete with a protective cover
   (d) An axe with a protective cover

(e) A hard helmet in a bright color
(f) Measuring equipment as required (e.g., tape measure or measuring rod)
(g) A whetstone
(h) A whistle

4.3.1.3
Bow Saws

In most forested countries the bow saw has replaced the crosscut saw for smaller trees, up to about 30 cm in diameter. Above this size, the crosscut saw performs better. The larger the tree or log being cut, the greater the problem of clearing the sawdust out of the cut. A long saw which permits clearing of the sawdust at both sides of the cut is the most efficient.

The bow saw consists of two end pieces: one with a handle, a center cross bar on one side of which is the saw blade, and on the other a rope which can be tightened by twisting to put tension in the saw blade.

The wooden frame (which can be readily made by hand) has been replaced in most countries with a tubular steel frame which is bent in a bow shape; hence, the name of the saw (Fig. 4.3). This bow-saw frame is bent in a manner which will put the tension in the blade. The tubular steel bow frame is light and easy to handle, but tends to lose its tension over time if the tension is not released when the saw is not in use. The wooden frame with the tensioning rope permits good control over the blade tension, both in and out of use.

Sharp teeth are important for efficient use. Saw blades with dull or improperly sharpened teeth, and with the teeth improperly set, result in an excessive expenditure of energy. One way of maintaining bow-saw efficiency is to use hardened-tooth blades. These blades are used until they are dull and then are discarded.

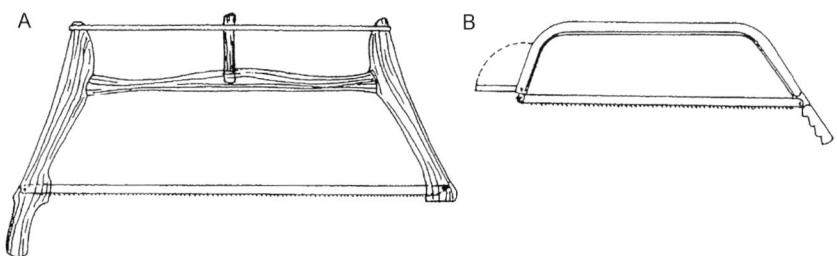

Fig. 4.3. Bow saws. **a** Wooden frame. **b** Metal frame

4.3.1.4
Power Saws

In natural forest, the power-saw team normally consists of two or three men: a feller, an assistant, and possibly a helper. The feller should be well trained in saw and saw-chain maintenance and sawing techniques. His assistant guides the feller during cutting and drives the wedges. He should also be qualified in operating the saw and should regularly relieve the feller. The helper clears the area around the tree. A smaller team is usually not favorable because too much time is lost for clearing work and the saw is not utilized as well. The following equipment is carried by the team:

1. Feller
   (a) The power saw (5–9 kW, guide bar 45–80 cm) with protective cover
   (b) The power-saw tool kit containing a multipurpose wrench, a double-head wrench, a round file and a spare chain (other maintenance tools for the saw chain to be kept at a road side shelter or the central workshop; mill saw file, slide gauge, depth gauge and vice)
   (c) A machete with a protective cover
   (d) A hard helmet in a bright color
   (e) A first-aid pocket kit
   (f) A whistle
2. Assistant
   (a) A sledge
   (b) Three wedges
   (c) An axe (weight about 1 kg, length of handle about 70–80 cm) with a protective cover
   (d) A whetstone
   (e) Measuring equipment as required (e.g., tape measure or measuring rod)
   (f) A machete with a protective cover
   (g) A hard helmet in a bright color
   (h) A whistle
3. Helper
   (a) A combined container for a fuel–oil mixture (5–10 L) and chain oil (2–3 L)
   (b) A machete with a protective cover
   (c) A hard helmet in a bright color
   (d) A whistle

In planted forests, the cutter usually works alone, although one or more helpers may be used to clean around the tree if the plantation has a lot of brush and to help in limbing. The power saw is approximately 55–60 cm³ and uses a bar of 40–45 cm.

4.3.2
Felling Methods

4.3.2.1
Axe Felling Smaller Trees

Axe felling is a job which requires both strength and skill. Precision is needed to let the tree properly break off from the stump and keep wood losses low. Smaller trees are normally felled by one worker (Fig. 4.4). The felling direction (Fig. 4.4, 1) usually corresponds to the lean of the tree. After determining the felling direction, the base of the tree and the retreat routes are cleared. The working space must allow clear movements of the axe which may easily cause severe injuries if its blow is deflected by branches or climbers. If buttresses are large, a platform must also be constructed.

The feller first makes the undercut (Fig. 4.4, 2) and ensures that it accurately faces the felling direction. The cut penetrates close to the center and opens at

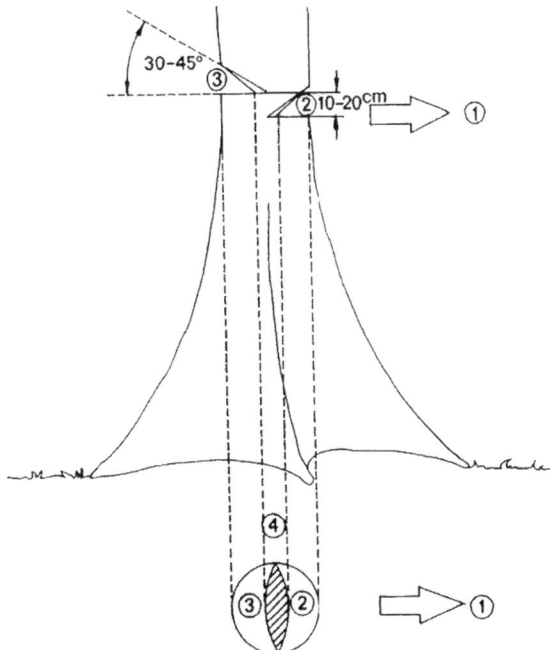

Fig. 4.4. Felling small trees by axe. For an explanation, see the text. (Courtesy FAO/ILO 1980)

an angle of 30– 45°. The backcut (Fig. 4.4, 3) is made from the opposite side at a level of 10–20 cm above the undercut. It opens also at an angle of 30–45°. Both undercut and backcut have about the same depth. For ease of work, it is preferable that their axis is slightly curved by removing more wood on the outer side leaving an elliptically shaped hinge (Fig. 4.4, 4). Once sufficient wood has been removed, the tree will bend over and the hinge will guide its fall into the felling direction.

In stormy weather, however, it is very difficult to maintain the intended felling direction. Under stormy weather conditions, felling should be stopped because of danger to the workers. Possibilities of felling trees purposefully in a direction other than the lean (directional felling) by means of axes are limited. By leaving a stronger hinge on one side, the tree will be pulled somewhat toward that side during the fall. A serious mistake is the felling of the tree by cutting it from all sides just as sharpening a pencil. By doing so the worker loses all control of the felling direction. Well-centered trees may fall to any side. The retreat cannot be planned properly and the accident risk is increased.

4.3.2.2
Axe Felling Larger Trees

Larger trees are normally felled by a team of two workers. Teams of more than two workers are usually less favorable because of the growing proportion of time lost by walking to and from the tree. Furthermore, delays are increased by workers hindering each other or waiting for each other. The accident risk is also increased.

Large trees will have to be felled almost exclusively according to their lean. Deviations through an asymmetric shape of the hinge (one side stronger than the other) become very difficult and cannot be expected in normal practical work (Fig. 4.5).

After determining the felling direction (Fig. 4.5, 1) the following preparations are made just as for small trees: clearance of the tree's base, clearance of retreat routes, and – if necessary – construction of a platform. Work then proceeds with undercutting up to a depth of about one third of the tree diameter (Fig. 4.5, 2). The undercut opens at an angle of about 45°. The backcut is carried out at a level of 20–30 cm above the level of the undercut. Work is facilitated if the backcut proceeds in three sections (Fig. 4.5, 3–5) penetrating to a depth of about one third of the diameter. Cutting of the last section (Fig. 4.5, 5) terminates the felling work and causes the tree to fall. The tree is guided by the hinge (Fig. 4.5, 6) into the felling direction. The backcut just above the undercut, opens at an angle of about 45°, allowing the axe to penetrate sufficiently deep into the tree without causing excessive waste of

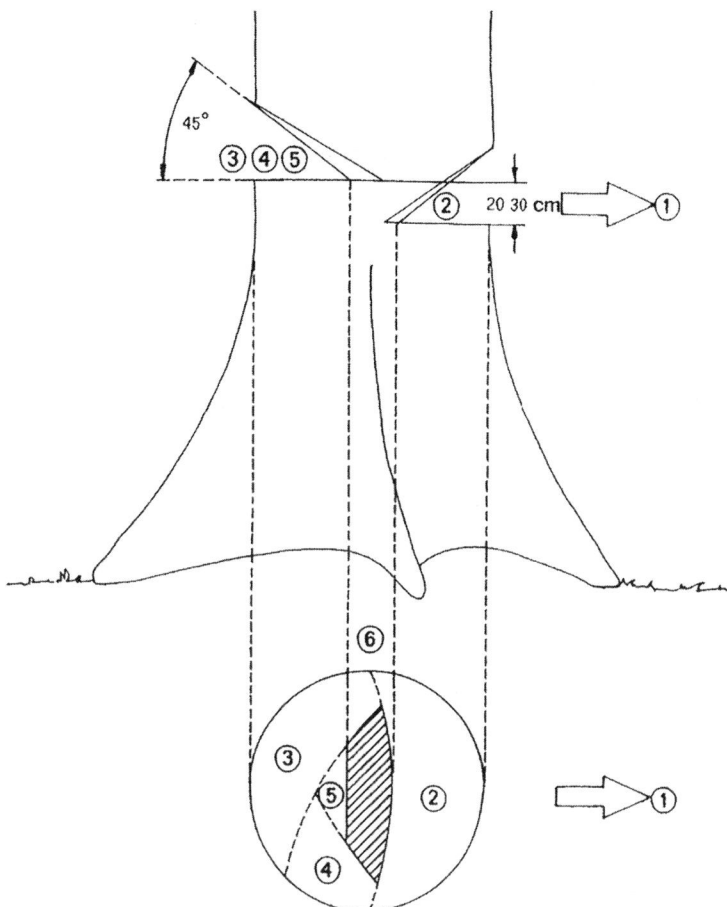

Fig. 4.5. Felling large trees by axe. For an explanation, see the text. (Courtesy FAO/ILO 1980)

wood and excessive effort. Axe felling of large trees is particularly dangerous in windy or stormy weather. Under stormy weather conditions, felling should be stopped because of danger to workers.

4.3.2.3
Felling Smaller Trees with Handsaws

Smaller trees up to approximately 60 cm diameter at the butt end are easily felled into the desired direction if they are cylindrical, sound, and straight, if their crown is extending evenly to all sides, if they are not tied up with other trees, if the base is easily accessible, and if the tree's fall is clear from obstruction (Fig. 4.6). If vines are interconnecting trees, the vines are often cut earlier to weaken the connections.

After determining the felling direction (Fig. 4.6, 1), the tools are placed behind the tree (Fig. 4.6, 2). Then the tree's base and retreat routes are cleared as described earlier. Work on the tree starts with the undercut (Fig. 4.6, 3), which should penetrate about one quarter to one third of the diameter into the tree. The undercut should be made close to the ground if the base of the tree is to be utilized or if high stumps would be obstructive for subsequent operations, such as skidding.

At first, the horizontal cut (Fig. 4.6, 4) is made with the saw. The oblique cut (Fig. 4.6, 5) can be made with the saw or the axe. The horizontal and oblique cut must meet in a straight line facing the felling direction of an angle of 90°. If stumps are liable to tear splinters from the tree, as can often be observed with softer woods, the undercut should be terminated by small lateral cuts (Fig. 4.6, 6) on both sides of the hinge (Fig. 4.6, 7).

The backcut (Fig. 4.6, 8) must also be made horizontally. It should be placed about 2.5–5 cm higher than the base of the undercut. The hinge is necessary to guide the tree during its fall. It should have the same width on both sides.

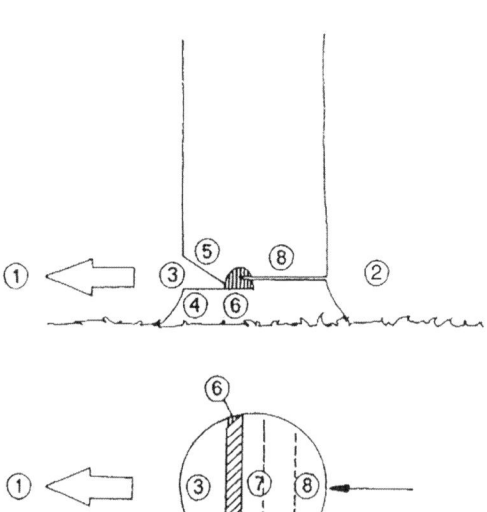

Fig. 4.6. Felling small trees by crosscut saw. For an explanation, see the text. (Courtesy FAO/ILO 1980)

4.3.2.4
Felling Larger Trees with Handsaws

Larger trees with more than 60-cm diameter at the butt end (Fig. 4.7) usually require a somewhat deeper undercut (Fig. 4.7, 1) penetrating about one quarter to one third of the diameter into the tree. The backcut (Fig. 4.7, 2) should be placed 10–20 cm higher than the base of the undercut. The hinge (Fig. 4.7, 3) must be stronger than with smaller trees. To facilitate the backcut, it should not be carried out parallel with the hinge. By alternating from one side to the other (Fig. 4.7, 4, 5) the sawing is less strenuous. Only the wood remaining in the center (Fig. 4.7, 6) is removed by sawing parallel with the hinge.

To make sure that the bases of the undercut and the backcut are cut at the intended level, it is advisable for the less experienced feller to mark the intended cuts with the machete before starting to saw.

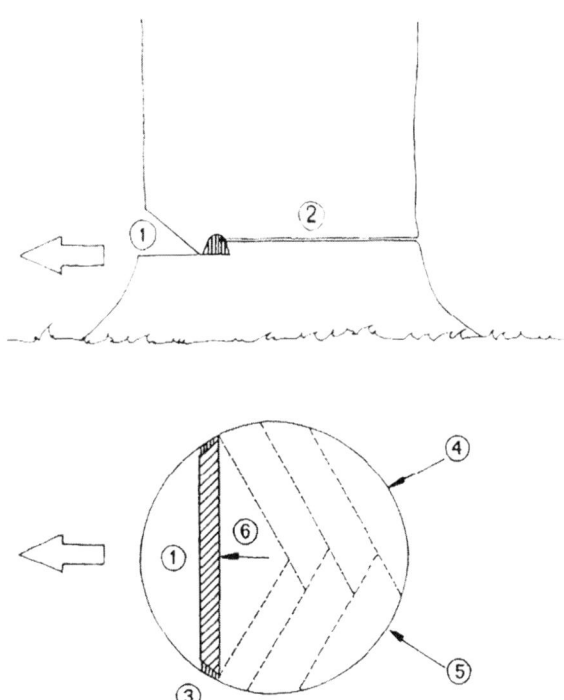

Fig. 4.7. Felling large trees by crosscut saw. For an explanation, see the text. (Courtesy FAO/ILO 1980)

4.3.2.5
Felling Smaller Trees with Power Saws

The helper leads the way to the tree, removing dense undergrowth and climbers from the path.

The felling direction is selected (Fig. 4.8, A) and the tools not needed immediately are placed at the opposite side (Fig. 4.8, B). While the feller prepares the saw (if necessary filling up the fuel–oil mixture and the chain oil, checking the tension of the chain, cleaning the air filter), the assistant and the helper clear the tree's base and the escape routes.

When cutting begins, the assistant joins the feller and guides him but keeps at least 2 m away from the saw to avoid being caught if it jumps away from the cut.

As in handsaw work the undercut (Fig. 4.8, C) must face the felling direction (Fig. 4.8, A) at a right angle and penetrate about one quarter to one third into the tree. The base of the undercut is made first and the oblique cut must neatly meet it in a straight line. Lateral cuts (Fig. 4.8, D) may additionally be required in soft wood.

The backcut (Fig. 4.8, E) must be horizontal and about 2.5–5.0 cm higher than the base of the undercut. If the saw's guide bar is sufficiently long, the saw is placed close to the hinge and the backcut is carried out in one continuous movement (Fig. 4.8, E). Special care must be taken, however, to leave the hinge sufficiently strong because otherwise control over the felling direction is lost.

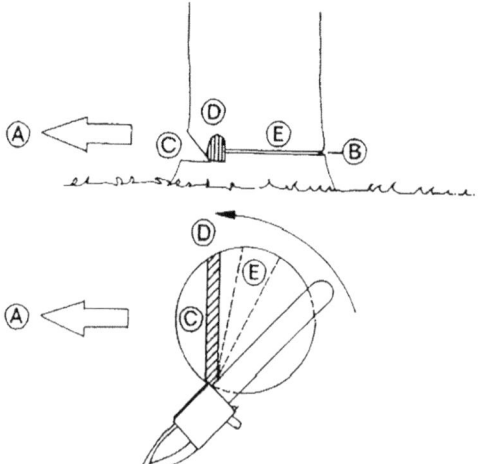

Fig. 4.8. Felling small trees by power saw. For an explanation, see the text. (Courtesy FAO/ILO 1980)

4.3.2.6
Felling Larger Trees with Power Saws

If the length of the guide bar is smaller than the diameter of the tree, the position of the saw must be shifted several times (Fig. 4.9) when performing the backcut. Alternatively, after the backcut has been completed, a boring cut can be made with the tip of the guide bar toward the center of the tree in order to remove that portion of the tree which is difficult to cut from the other side (Fig. 4.10). Care must be taken to ensure that hinge remains strong enough on the two sides of the cut.

In felling larger trees suspected of being hollow, it is recommended that the feller first tests the tree for hollowness by inserting the saw blade in a vertical plane into the stem.

4.3.2.7
Felling Trees with Plank Buttresses

Plank buttresses are common in large tropical trees. Large tropical trees with plank buttresses often become cylindrical at heights of 3–5 m above the ground. With axes, a platform is often built to get above the main part of the buttresses. With power saws and handsaws, the trees are felled at a height of about 80 cm above ground level. At this height the tree will have sufficient center wood to remain stable once the buttresses have been cut so that it can be felled in the desired direction.

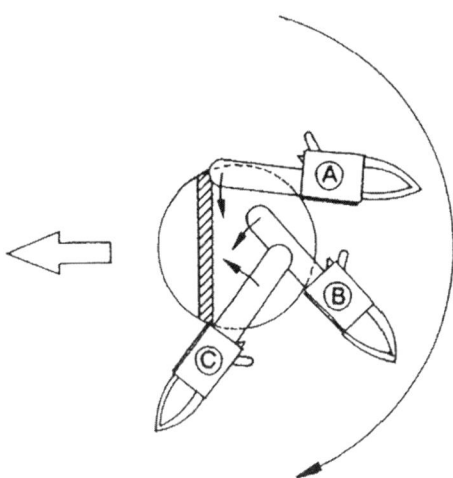

Fig. 4.9. Felling large trees by power saw. (Courtesy FAO/ILO 1980)

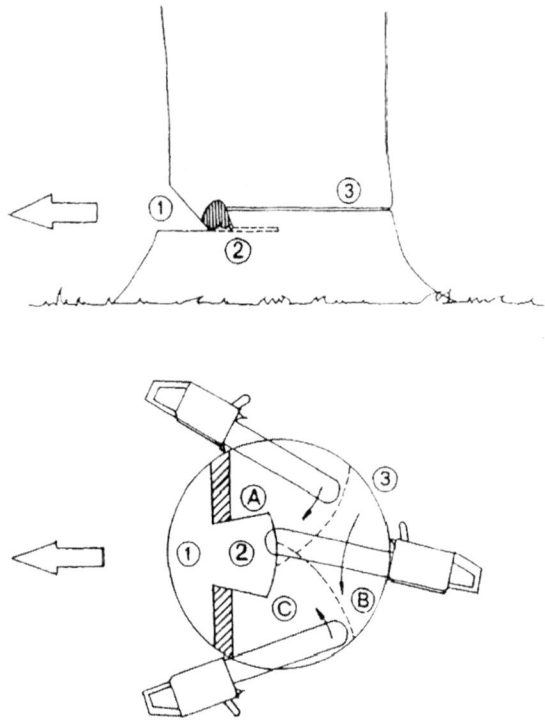

Fig. 4.10. Felling large trees by power saw using alternative techniques *Numbers* indicate the sequence of cuts with the chainsaw. (Courtesy FAO/ILO 1980)

Techniques for felling of trees with large buttresses with handsaws and power saws are similar; however, power saws have the advantage of being easier to operate in a limited space.

After the felling direction (Fig. 4.11, A) has been selected, the undercut (Fig. 4.11, B) is made to a depth approximately one third of the diameter. The backcut (Fig. 4.11, C–E) should be made about 20 cm higher than the base of the undercut. It begins on the lateral buttresses (Fig. 4.11, C, D) and is terminated on the backward one (Fig. 4.11, E).

4.3.3
Bucking

Once a tree has been felled, work should continue by bucking (crosscutting) the tree by the same team that felled the tree. Log lengths will depend upon tree size and the skidding system. Workers involved in bucking must have basic knowledge in tree grading and measuring. An alternative is to concentrate

Fig. 4.11. Felling trees with buttresses. (Courtesy FAO/ILO 1980)

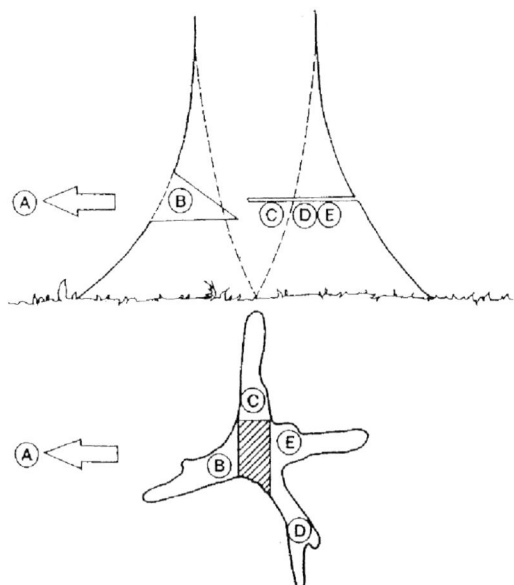

bucking at the landing where logs can be placed free of tension and measurements can be made by personnel more highly skilled in log grading. In this case only the crown of the tree needs to be cut off at the felling site. However, the skidding equipment must be sufficiently powerful to pull the tree length stem to the landing.

4.3.4
Felling and Bucking Production

4.3.4.1
Natural Forests

The estimated production per hour and per day in felling and bucking can be read from the nomograph in Fig. 4.12. The selected production factors are tree diameter (diameter at breast height, DBH, above possible buttress), number of logs per tree, and effective working time.

The number of hours effectively worked per day is one of the most significant factors affecting production. It is also one of the most difficult to establish. Experience in tropical forests shows that the level of six effective hours is seldom reached in felling mainly owing to physical exhaustion in the hot climate. In this operation, effective working time includes walking to the trees, handling of tools, preparation at the stump site, sawing, and minor items, but excludes long pauses and delay time.

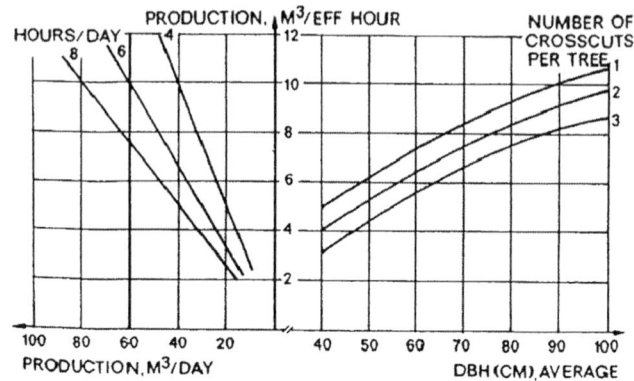

Fig. 4.12. Felling and bucking with power-saw (one operator and one helper). Hours per day refer to number of effective hours per day. *DBH,* diameter at breast height. (Courtesy FAO/ILO 1980)

Table 4.2. Felling and bucking production in the stump area

Method	Normal daily crew production (m³)	Number of workers in crew
Axe for felling and bucking	3–15	1
Axe for felling, two-man saw for bucking	25–30	2
Two-man crosscut saw for felling and bucking	15–30	2
Power saw	30–70	2

Table 4.2 presents information on methods and production for stump area operations typical of high tropical forest conditions.

Figure 4.12 is read in a counterclockwise direction, starting on the horizontal axis at the right (average DBH). From there, read up to the expected number of logs (bucks) per tree and then left to the production per hour (on the vertical axis). The expected production per day or shift is then read on the horizontal axis at the left after establishing the expected number of effective working hours.

The following correction factors that may have to be applied to the production estimates from Fig. 4.12:

- For stand and terrain conditions the values given in Table 4.3 should be applied.
- When platforms are built for felling above the buttress or buttswell, reduce production by 20% by multiplying the production figure by 0.80.
- If the expected volume should deviate from the volume assumed in designing Fig. 4.13, for example, if an 80-cm DBH tree is expected to give 7 m³ of

Table 4.3. Correction factors for stand and terrain conditions

Terrain and vegetation on the felling site	Form and quality of trees Tall and well-formed trees and/or little damage	Normal length and form and/or normal damage	Short and badly formed trees and/or severe damage
Steep terrain (more than 40%) or swampy ground and/or severe felling obstacles (underbrush etc.)	0.9	0.8	0.6
Average – normal conditions	1.2	1.0	0.8
Smooth or undulating terrain, well-drained soils, no severe underbrush or other felling obstacles	1.5	1.2	1.0

merchantable timber instead of 6 m³, a correction must be made. This is done by multiplying the production figure by the quotient between expected and assumed volumes – in the above case by 7/6.

- It is assumed in the nomograph that the work is carried out with one-man power saws being used by one operator and one helper. This is not to say that two men per saw is always the case. A third worker may be attached to the crew to help with supplying fuel, etc. If so, the output will increase by 5–15%. This might be worthwhile in areas where the harvested volume is low. In such cases the second helper acts as a scout and locates the trees to be felled, thus decreasing walking and preparation time for the saw operator.
- It is assumed in the nomograph that bucking is made at the felling site before skidding. If additional bucks are made at the roadside, this work operation can be assessed approximately by taking the difference between the readings in Fig. 4.12 for the number of bucks at the felling site and at the roadside, respectively, in terms of their time consumption.

Examples: Estimate felling and bucking productivity by power saw in natural forest for a 5 hour effective day with an average DBH of 65 cm for (a) one crosscut per tree at the felling site, (b) three crosscuts per tree at the felling site, and (c) bucking productivity for making two crosscuts at roadside instead of at the felling site.

- Reading for three bucks gives 30 m³/day, which equals 0.033 day/m³
- Reading for one buck gives 40 m³/day, which equals 0.025 day/m³
- Time consumption for two bucks at the roadside is 0.008 day/m³ or a production of 125 m³/day

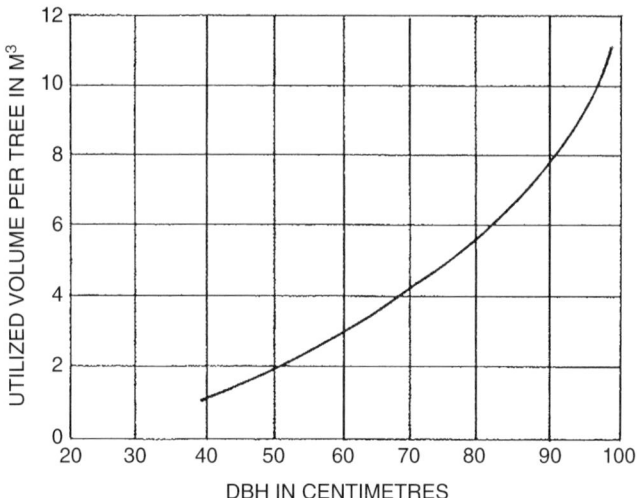

Fig. 4.13. Utilized volume per tree. (Courtesy FAO 1976)

4.3.4.2
Planted Forests

Owing to the smaller size and homogeneity of wood in planted forests there are more options for handling the wood than in natural forests. There are many choices in the bucking and limbing activities. One of the main features in the design of a logging system is the prescription of the form and position of the wood when it is passed from one main phase of logging to the next. Trees may or may not be delimbed, topped, bucked, debarked (peeled), bundled, or piled in one form or another. Some of these jobs may be only partially performed; for example, rough delimbing, bucking in multiple log lengths, strip barking.

To present production data for so many options is not feasible, but two main cutting procedures are covered: cutting in tree lengths and cutting shortwood, in both cases excluding debarking, for which separate production data are given. Some indications are also given of the influence on production of additions or simplifications within the two procedures. The output in cutting is extremely variable and depends for each method and type of equipment on two main sets of variables: physical (tree and stand parameters and other environmental factors) and human ability and the skill of the labor force.

Experience has shown that the influence on production of the physical variables is rather conformative and stable in relative terms. The same variable exerts approximately the same relative influence – say increase or decrease of 10% in time consumption – independently of the standard of the labor force.

Consequently, it is recommended that an estimation of output in cutting – or in any manual logging job – should be made in three steps:

1. Step 1. Establish the output (or time consumption) as related to the physical variables, assuming the work is performed by a skilled and able labor force working on a piece-rate basis. In principle, this estimate will give the highest level of production that can be expected.
2. Step 2. Compile the characteristics of the labor force.
3. Step 3. Adjust the production figure from step 1 with regard to the influence of the human variables. In other words, analyze in what respect the actual labor force differs from a skilled and able one and estimate the reduction in production likely to result from such differences. Production in the cutting of tree lengths and shortwood is given in two basic curves for one-man work, using power saws for felling, delimbing, and bucking. The production is related to the mean DBH of the trees and for tree crowns at three different heights:
   (a) Short: at 25% of the tree height
   (b) Average: at 50% of the tree height
   (c) Long: at 80% of the tree height

To derive the crown length, add half the length of the bole with dead branches to the length of the green crown. Production as read on these curves should be adjusted for three sets of variables: method and equipment, physical variables, and human variables.

4.3.4.3
Cutting Tree Lengths

Production in trees per day is given in Fig. 4.14. Production as read on the curve should be adjusted (reduced) for the three sets of variables, as follows:

1. Adjustments for method and equipment
   (a) Manual saw and axe instead of a power saw: reduce production by 8% for DBH of 10–20 cm, 20% for DBH of 20-40 cm, and 30% for DBH greater than 40 cm.
   (b) Delimbing with an axe instead of a power saw: reduce production by 15% for a DBH over 15–20 cm.
   (c) Rough delimbing allowing stubs to remain: increase production by 15%.
   (d) Two-man team instead of one man: increase production by 70%.

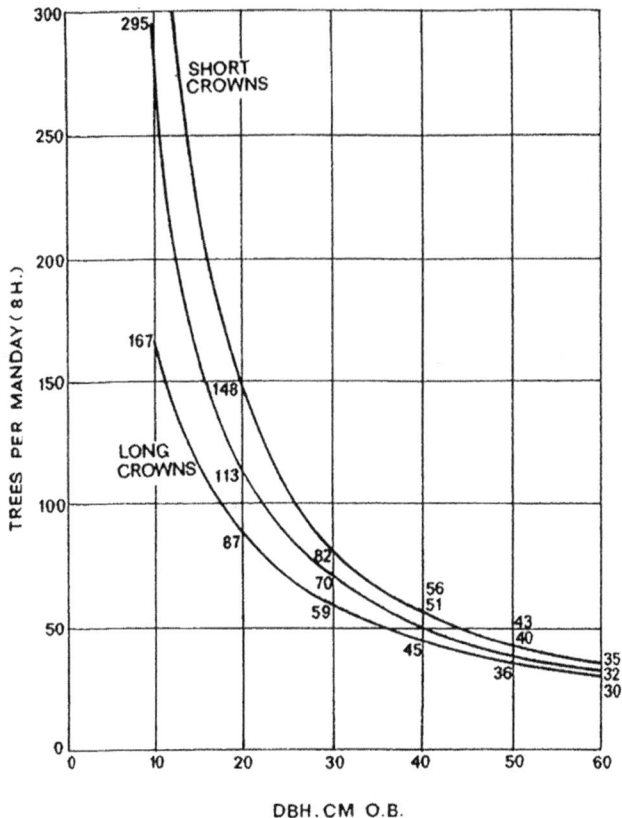

Fig. 4.14. Production for cutting tree lengths. This chart is applicable under the following conditions: one man using a power saw for felling, delimbing, and bucking; no debarking; minimum top diameter about 7 cm for trees up to DBH of 30 cm and about 10 cm for larger trees; easy terrain; temperate climate. (Courtesy FAO 1976)

2. Adjustments for physical variables
   (a) Adjustment for terrain: reduce production by 5% for undulating terrain (slopes greater than 20%), 15% for steep terrain (slopes greater than 35%), and 30% for very steep terrain (slopes greater than 50%).
   (b) Make adjustments for any extreme condition such as very dense underbrush, extremely thick limbs, much cull wood, or crooked boles. In general no reduction of more than 10% should be made for each factor.
   (c) Make adjustments for climatic conditions.

3. Adjustments for human variables
   (a) The production curves apply to forest workers weighing on average about 73 kg. For lighter workers the following productivity adjustments are suggested:

Average body weight (kg)	Productivity multiplier Axe, bow saw, heavy manual work	Power saw
70	0.96	0.98
65	0.90	0.95
60	0.83	0.91
55	0.76	0.86
50	0.69	0.82

   (b) For new workers just learning the job, reduce the production estimate by as much as one third. Experience shows that trained workers over a 2-year period can improve their production as much as 50% over workers with less than 6 months' experience.
   (c) Consider the season and location of the workplace. High temperature, humidity, and elevation all reduce productivity.

Example of cutting tree lengths: A cutter weighing 60 kg is felling and delimbing using a power saw on steep terrain. The average stand diameter is 20 cm and the trees have medium length crowns.

- Step 1. From Fig. 4.14, the production is 113 trees per 8-h day.
- Step 2. Reduce by 15% for steep terrain, 0.85 × 113 = 96 trees per 8-h day.
- Step 3. Reduce by 9% for body weight, 0.91 × 96 = 87 trees per 8-h day.

4.3.4.4
Cutting Shortwood

Production in trees per day is given in Fig. 4.15. Production as read on the curve should be adjusted (reduced for the three sets of variables) as follows:

1. Adjustments for method and equipment
   (a) Manual saw and axe instead of a power saw: reduce production by 10% for a DBH of 10–20 cm, 20% for a DBH of 20–40 cm, and 30% for a DBH greater than 40 cm.
   (b) Delimbing with an axe instead of a power saw: reduce production by 15% for a DBH over 15–20 cm.

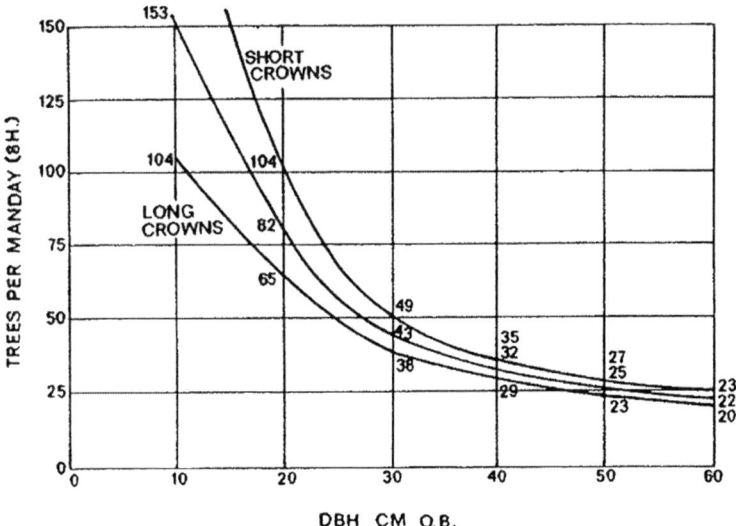

Fig. 4.15. Production for cutting shortwood. This chart is applicable under the following conditions: one man using a power saw for felling, delimbing, and bucking; no debarking; minimum top diameter abut 7 cm for trees up to a DBH of 30 cm and about 10 cm for larger trees; saw logs cut in 4–5-m lengths, pulpwood not shorter than 2.5 m; pulpwood bundles in piles containing less than 0.3 m³; easy terrain; temperate climate. (Courtesy FAO 1976)

(c) Rough delimbing allowing stubs to remain: increase production by 15%.

(d) Pulpwood cut in lengths shorter than 2.5 m: reduce production by 15% for a DBH less than 20 cm and by 5% for a DBH greater than 20 cm.

(e) Bunching of pulpwood in piles containing more than 0.3 m³: reduce production by 10% for a DBH less than 20 cm.

2. Adjustment for physical variables

(a) Adjustment for terrain: reduce production by 5% for undulating terrain (slopes greater than 20%), 15% for steep terrain (slopes greater than 35%), and 24% for very steep terrain (slopes greater than 50%).

(b) Make adjustments for any extreme condition such as very dense underbrush, extremely thick limbs, much cull wood, or very crooked boles. In general no reduction of more than 10% should be made for each factor.

3. Adjustments for human variables: same as for longwood

Example of cutting shortwood: A cutter weighing 60 kg is felling, delimbing, and bucking using a power saw on flat terrain. The average stand diameter is 20 cm, the trees have long crowns, the stems are crooked, and the log length is 2.5 m.

- Step 1. From Fig. 4.15, the production is 65 trees per 8-h day.
- Step 2. Reduce by 10% for crooked stems, $0.9 \times 65 = 59$ trees per 8-h day
- Step 3. Reduce by 9% for body weight, $0.91 \times 59 = 53$ trees per 8-h day.

4.3.4.5
Debarking

The volume of production in forest debarking depends largely on log diameter and the properties of the bark. The bark surface per unit volume of wood is correlated to the log diameter as illustrated below:

Log diameter (cm)	4	8	10	15	20	25	30
Bark surface (m^2/m^3)	100	50	40	27	20	16	13

Consequently, in both manual and mechanical debarking (excluding drum debarking at mills), production decreases sharply with decreasing log diameter. The properties of the bark must also be taken into consideration. Bark can be most easily removed from the wood during the sap season. The drier the wood, the more difficult the debarking. Thick bark and irregularities on the surface of the wood make debarking more difficult.

Several tools are used for manual debarking, especially in the sap season in peeling certain wood species, particularly hardwood. The barking spud is probably the most efficient as soon as the bark does not peel off very easily. As a spud will remove a narrower strip of bark on small logs than on logs of larger diameter, production tends to decrease more for small wood than implied by the ratio of wood per unit area of bark. A high output in debarking of softwood by a trained worker with a barking spud under favorable conditions is 50 m^2 of bark surface removed in one effective work hour. This is reduced by half for a log diameter of 8–10 cm. A reduction to half or less may easily follow if after felling the wood is stored before being debarked and the logs are allowed to dry. Even a few hours' delay may under certain conditions cause significant differences in some species, such as eucalyptus. It should also be noted that manual debarking is a heavy monotonous job and a sustained production of 50 m^2 is seldom attained. Figure 4.16 can be used as a guide where local experience is not available.

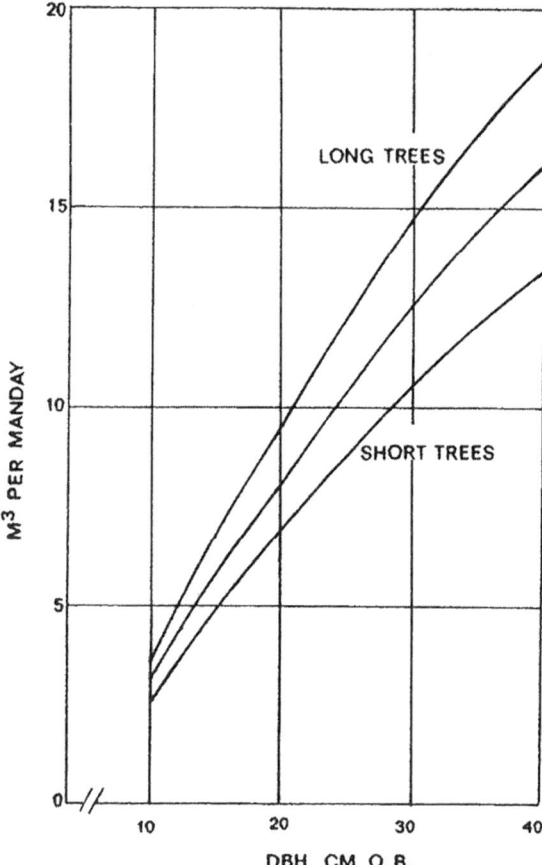

Fig. 4.16. Production for manual debarking. This chart is applicable under the following conditions: one man working with barking spud; debarking performed immediately after felling; easy terrain. (Courtesy FAO 1976)

The graph takes into account that short trees, in addition to high taper, generally also have thicker bark and more knots and irregularities in the surface of the wood. Production as read on the curve should be adjusted as follows:

- Reduce estimates by 30–40% for conditions in which the bark adheres markedly to the wood, for example, when felled logs have been allowed to dry.
- Reduce estimate by 10–15% for steep and very steep terrain when debarking is done in the stump area.
- Physical work capacity is a limiting factor of production in manual debarking, so make adjustments for human variables as suggested in cutting longwood.

4.4
Mechanized Felling and Delimbing

4.4.1
Mechanized Felling

Mechanized felling is the cutting of trees by one of a variety of mechanical cutting heads attached to and powered by a rubber-tired or tracked carrier. Mechanized felling is increasingly being used in planted forests. It has the advantages over manual felling of:

- Increased worker safety by removing the worker from the forest floor
- Much higher production rates per man-hour
- Less waste through cutting of lower stumps
- Lower breakage through better tree placement
- Increased efficiency of skidding through bunching of smaller trees for grapple skidding

Mechanized felling has some disadvantages relative to manual felling:

- Higher capital investment
- Potential for more ground disturbance
- More limited by terrain, including obstacles, ground roughness, steepness, and moisture conditions

The cutting heads are usually hydraulically powered and include various types of shears, chain-and-bar saw heads, and circular saw heads. Circular saw heads are further subdivided into continuous and intermittent. Continuous saw heads have inertial disks revolving continuously during operation. Intermittent saw heads are activated and revolve only when the felling machine is at the tree to be cut. The cutting heads are mounted on a variety of carriers, including both rubber-tired and tracked front-end loaders, excavators, and forwarders. Sometimes the felling is carried out as a separate activity (single function) and the machine is called a feller-buncher. Other times the felling is combined with delimbing, bucking, and topping the tree (multiple functions) and the machine is called a harvester.

Maximum tree diameters for shear-type cutting heads are generally under 55 cm in softwoods and less in hardwoods. Some saw-head models can cut trees up to 75 cm in diameter.

Feller-bunchers can substantially increase skidding productivity by cutting and piling (bunching) the trees for skidding by grapple skidders. Often the cutting heads on feller-bunchers are combined with grapplelike arms to accumulate two to six small trees into a bunch before dropping the trees to the ground.

Carriers are divided into two basic types: tree-to-tree, and limited area (Fig. 4.17). Tree-to-tree carriers must drive to each tree prior to felling. Limited-area carriers have the cutting head mounted on a hydraulically powered articulated swivel-boom which can reach out from a stationary location to cut several trees before moving on. Limited-area carriers, particularly on tracks, have a relative advantage over rubber-tired carriers on slopes and low-bearing ground surfaces. They also have an advantage if it is necessary to reach around obstacles such as large tree trunks left from natural forest.

Production rates for feller-bunchers depend mainly upon the type of saw head, tree diameter, and number of trees per acre. Production varies from one to three trees per minute and can be as high as 200 trees per hour (Fig. 4.18).

4.4.2
Mechanized Delimbing

Mechanized delimbing in planted forests can be done with a variety of techniques, including gate delimbing, mechanical delimbing knives, and chain flails (Table 4.4).

With a gate delimber, a tree is delimbed by pushing the tree through a locally made iron-lattice framework known as a delimbing gate. The gate is often supported by chaining it between two standing trees. The tree being skidded is pushed through the gate by backing up a grapple skidder and pushing the tree

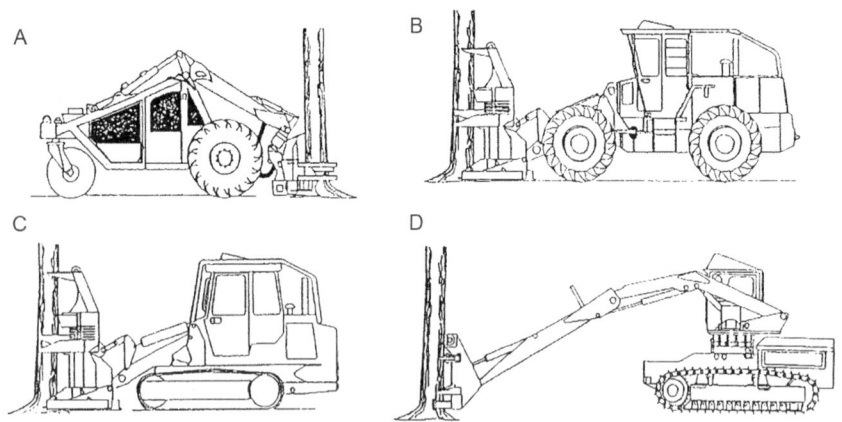

Fig. 4.17. Carriers for tree-to-tree feller-bunchers: **a** Three-wheel. **b** Rubber-tired loader. **c** Track loader. **d** Track carrier for a swingboom, limited-area feller-buncher

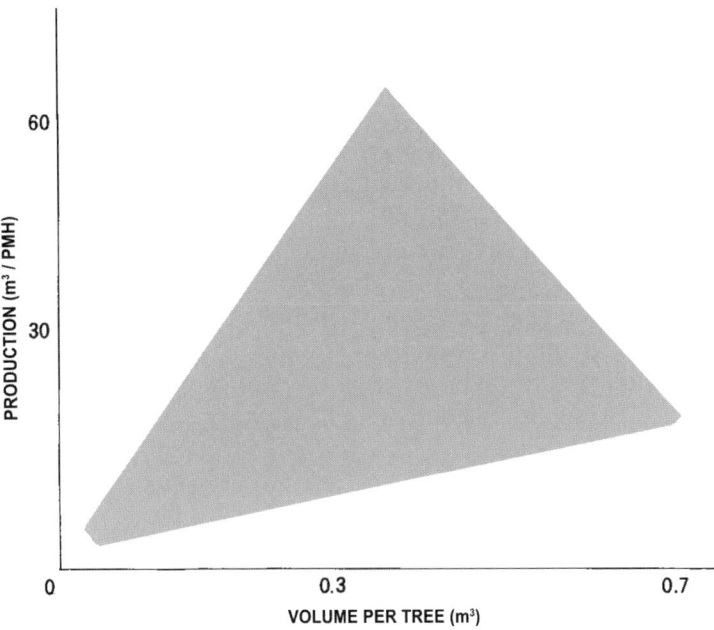

Fig. 4.18. Range of production observations from various studies of feller-bunchers as a function of tree size

Table 4.4. Comparison of delimbing options

Method	Advantage	Disadvantage
Delimbing gate	One man skids and delimbs Usually used in tree length skidding	Reduces skidder productivity Increases log breakage Concentrates branches
Stroke-boom	Delimbs, bucks at the roadside Permits tree length skidding Has potential to optimize tree bucking since the entire tree or most of it is measured during delimbing	Requires separate machine Concentrates tops and branches at landing
Felling/processing head	Same machine combines felling, delimbing, and bucking Limbs and tops placed in skid trail so rutting is reduced Shortwood permits forwarding	Productivity is lower than for a feller-buncher Reduces log bucking options at the mill
Chain flail	Flail can both delimb and debark	Increased breakage. Breakage not easily recoverable if flail is not at the mill

between iron bars. As the tree is pushed through the gate, its branches are broken off. The method works well with conifers having brittle branches. Mechanical knives are of two types: with one, the knives are pulled along the tree, with the other, the tree is pulled through the knives. Mechanical delimbers which pull knives along the tree are called stroke-delimbers (Fig. 4.19). Stroke-delimbers are normally used at the landing and are mounted on either rubber-tired or tracked carriers.

Delimbers which pull the tree through the knives are usually, but not always, part of a processing head which also fells and bucks the tree into logs. Processing heads can be mounted on rubber-tired or tracked carriers.

Chain flails use rotating chain links to snap off branches. Chain flails can be mounted on a rubber-tired carrier which runs over the logs to be delimbed or delimbed by a flail mounted on a semistationary platform at the landing.

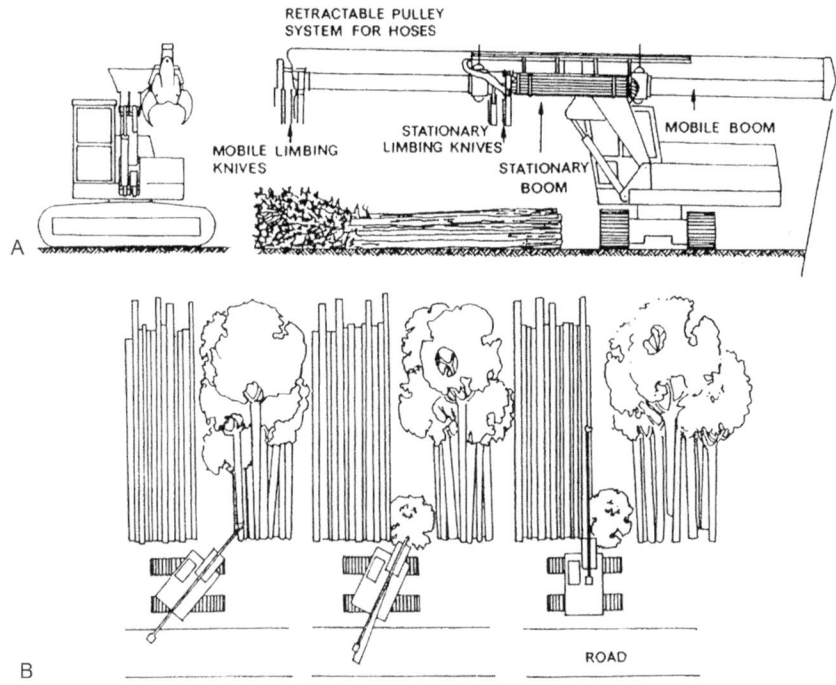

Fig. 4.19. Track mounted stroke-boom delimber. **a** Front and side views. **b** Picking up tree, delimbing, topping, and stacking

4.4.3
Harvesters

Harvesters are machines which fell, delimb, and buck tree stems. Harvesters are divided into two groups. Two-grip harvesters first sever the tree from the stump with a crane-mounted felling head and then transfer it for further processing to a separate mechanism mounted on the base carrier. One-grip (single-grip) harvesters (Fig 4.20) use a crane-mounted device for felling, and subsequent delimbing, measuring, and bucking. Single-grip harvesters are the most common and are more economical than two-grip harvesters for trees less than 0.5 m³. The base carrier for the harvester can be either track or rubber-tired. All harvester heads are hydraulically driven. Harvesters operate in both selective and clearcuts. Harvesters can be equipped with lights and often work two shifts in order to more effectively use the capital investment.

With the single-grip harvester, the tree is severed with a hydraulically driven chainsaw within the head, the tree is then rotated parallel to the ground and is then pulled through delimbing knives by hydraulically powered rollers. Potentiometers in the knives measure stem diameter and a measuring wheel records the length and these supply information to assist in the bucking decisions. Log lengths are controlled by preprogrammed instructions from an on-board computer. Volume is calculated automatically on the basis of log

Fig. 4.20. Single-grip harvester rotating tree after felling to begin delimbing, measuring, and bucking process

lengths and diameters, providing a record of production on the computer. Some harvesters can transmit this information digitally. Harvesters work on slopes up to 40%, but productivity decreases on slopes greater than 30%. Processing is done ahead of the machine, creating a mat of limbs and tops over which both the harvester and the forwarder pass. The reach of the harvester depends on machine size but is generally 7–10 m. Optimal operation of the harvester requires straight stems and small branches. Most delimbers will work well when the branch size is less than 5 cm. Larger branches will require heads with more powerful delimbing force. Log measurement can be problematic during periods when the bark is loose or in species with nodal swelling. Productivity is heavily influenced by tree volume (Fig. 4.21).

4.5
Maximizing Tree Value

4.5.1
Natural Forests

In natural forests obtaining maximum tree value centers upon (1) proper identification of the tree to be felled, (2) identifying a safe bed where the tree bole will not be broken, (3) placing an appropriate undercut and backcut that does not split the tree trunk and provides a clean hinge, (4) bucking the tree according to log specifications that maximize tree utilization and value, (5) minimizing log loss during the skidding operations, (6) pricing policies

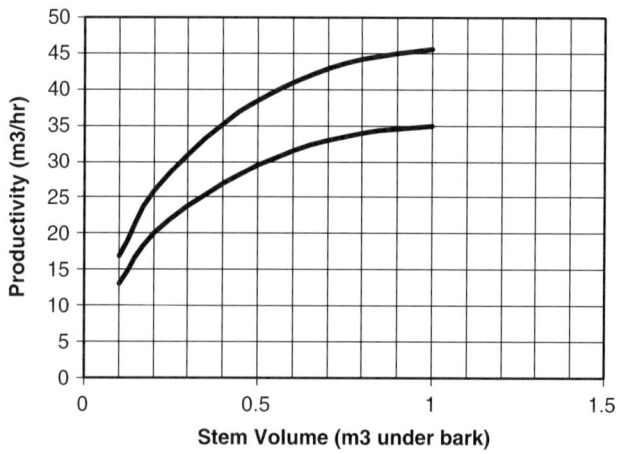

Fig. 4.21. Estimated range of production per effective hour for a single-grip harvester with an experienced operator clear cutting on level ground in planted forest

that encourage tree utilization, and (7) supervision and enforcement of administrative procedures to curb poor utilization. Under some concession policies, logging contractors are required to pay for only what is skidded to the landing or loaded on trucks. This policy can result in poor utilization and value recovery since poorly utilized trees and lost logs are not penalized. Similarly, policies that charge a single price per cubic meter regardless of log quality discourage good utilization because low-value logs must be paid for at the same price as high-value logs and will be left behind. Pricing policies should also recognize logging costs. Logs in difficult areas or requiring long extraction distances should be priced appropriately to encourage utilization. Average log pricing will encourage logs only in the easiest locations to be utilized. Proper supervision is necessary to identify poor utilization practices and to enforce administrative procedures.

4.5.2
Planted Forests

Log buyers often have choices of suppliers. Grading and sorting during harvesting determines the quality of the raw materials entering the mills they supply, the grade out-turn, productivity, and profitability of those mills. The capability of a harvester to provide logs that meet the customer's needs will affect the ability to make a sale, the maintenance or improvement of market share, and the price a buyer is prepared to pay. Logs that do not meet specification are likely to be rejected, incurring the additional handling and transportation costs for their removal or the costs of remanufacturing to meet specifications.

Between standing tree and delivery of logs to the customer there are many ways that value may be lost along the supply chain. Considerable financial losses can occur when volume is sacrificed or the wrong log grade selections are made from each tree. Harvesting managers should not accept losing value as a part of doing business.

Practical suggestions for achieving maximum value, and minimizing costs of poor quality and wastage, are provided in Table 4.5. Monitoring quality and value recovery is vital for the success of a forest enterprise – as the old adage goes, "if you don't measure it, you can't manage it." To maximize the value of the resource for the forest owner a commitment to managing quality and achieving high levels of value recovery needs to pervade all levels of a forest enterprise.

Table 4.5. Practical suggestions for achieving maximum value from harvested stands (From Hammond 1995)

Activity	Item	Desired result	Consequences if ignored
Felling	Felling direction	Parallel stems aligned for extraction	Excessive breakage, volume and value loss
	Stump heights	Low stumps	Loss of pruned log volume and value
	Mechanical felling heads	Complete cut or multiple cuts if diameter is too large	Incomplete cut leads to end splitting and slabbing of butt log
	Wing cuts – manual felling	Used frequently and correctly	Slabbing and butt log damage
	Wedges – manual felling	Directional control	Excessive breakage
	Pruned zone identification	Mark just below first branch	Pruned log quality errors
Delimbing	Manual delimbing	Branches trimmed flush with stem	Logs rejected because of long stubs
	Machine-feed rollers	No impact on tree stem	Spikes and feed mechanisms damage logs and increase risk of fungal infection
	Mechanical delimbing knives	Well maintained	Log downgraded through branches being pulled out rather than cut cleanly
Tree extraction	Stem handling	Minimal damage	Downgrading of logs owing to grapples or extraction breakage
	Fleeting and alignment of stems	No damage	Blading damage to logs
	Stem positioning on landing	Open with butts apart	Difficult to inspect log quality
Manual processing	Log specifications	Carried on site by logmaker	Logs do not meet specifications. Misunderstanding of specifications
		Understood fully and consistently	Different interpretations of grading rules
	Log product cutting priority	Formal control of cutting instructions with guidelines for priority cuts	Value loss owing to log making not matching production priorities

Table 4.5. Practical suggestions for achieving maximum value from harvested stands (From Hammond 1995)—Cont'd

Activity	Item	Desired result	Consequences if ignored
	Tools (tapes, calipers, etc.)	Calibrated and used correctly	Logs do not meet specification
	Cutting	Straight	Angled or end splits owing to tension breaks
	Sloppy cut	Minimum cut	Loss of pruned log volume
	Location	Free of mud, with adequate room to safely accommodate machines and people	Quality errors and value loss owing to inability to inspect logs safely
	Production	Logmaker-paced production system with sufficient time for inspection and decision making	Machine-paced operation resulting in quality errors and high value loss
Mechanical processing	Cuts	Minimal end splits	End splits can be caused by stems that are cut under tension, poorly maintained or sharpened saws, cutting large diameter logs too slowly
	Diameter and length measurement	Well-designed sensors that are checked and calibrated regularly	Log outside diameter and length tolerances can be caused by faults in measurement devices owing to design or maintenance or slippage of logs in feed mechanisms
	Bark thickness	Accurate underbark diameter prediction	Value loss and logs not meeting minimum or maximum diameter specifications
	Assessment of log quality attributes	Adequate ergonomic design that assists inspection of stem quality features	Logs incorrectly graded or rejected owing to poor stem visibility to operator
Sorting	Branding of logs	Clear identification of grade, crew, and date of production	Incorrect sorting and poor traceability of products

(Continued)

Table 4.5. Practical suggestions for achieving maximum value from harvested stands (From Hammond 1995)—Cont'd

Activity	Item	Desired result	Consequences if ignored
Storage	Stocks	Daily monitoring	Downgrading owing to decay
	Log piles	Logs stored out of ground contact with adequate circulation of air	Downgrading owing to decay
Harvest planning	Stand inventory	Correct product predictions	Oversupply or undersupply of products to customers, owing to incorrect use of sampling procedures
	Stand allocation	Log products that match the forest resource with markets	Suboptimal value obtained from forest owing to inappropriate product mix selection
Management	Documentation	Correct	Incorrect invoicing and payment, mistakes in deliveries
	Dispatch	Delivery to customer on time	Late or early delivery can lead to mill stockpiles that are too high or too low
	Company quality policy and procedures	Documented and well publicized	Disparate goals, procedures, and product standards
	Reporting and auditing	Periodic reports to manager on quality and value achievement	Lack of data for management control or decision making
	Personnel selection	Select people with attributes and skills suited to their job	Unskilled and unmotivated work force with high turnover and accidents, low productivity, and poor quality
	Training programs	In place with regular refreshers, recognition, and rewards for qualifications	
	Liaison with customers	Regular feedback mechanisms	Poor understanding of customer needs

Skidding by animals is limited to smaller logs from natural forest and pulp-wood and thinnings from planted forests. In planning for animal skidding, careful consideration must be given to the capacities of the animals and their maintenance requirements. When animals are used for skidding purposes, feeder roads should be more closely spaced and grades must be more limited than when skidders are employed. Horses are not suited for skidding work in hot climates for physiological reasons. Furthermore, their purchase cost is higher and their life span shorter; their feed must be more selective and, generally speaking, their maintenance cost is higher.

Mules, donkeys, oxen, water buffalo, and elephants are suitable for work in the tropics. Logs can be skidded or forwarded using skidding chains, sulkeys, or carts. When chain skidding, the use of a skidding pan or sledge will reduce skidding resistance, allowing heavier loads to be pulled for longer distances. Proper harnesses are essential to avoid discomfort and injury to the animals (Fig. 5.1).

5.1
Mules

Mules, when fed properly three times a day, are capable of working steadily 5 days a week, although they may refuse to work everyday. A 20% reserve of animals is usually required to replace those sick or injured. Mules are usually worked singly. Maximum skidding distance is about 140 m within grade limitations of 30% downhill and 15–17% uphill. On soft soils, a maximum downhill gradient of 25% is preferred for safety reasons. Skid trail gradient is an important factor affecting load size (Fig. 5.2). Production is so reduced when skidding uphill that the practice, while possible, is hardly worthwhile on all but minor grades. When skidding is downhill, the size of the load tends to level off as the grade increases and seldom reaches 0.2 m^3. The skidding chain should be kept short to provide a lifting effect to the forward end of the logs when skidding is uphill and on the level, and lengthened on the steeper downgrades

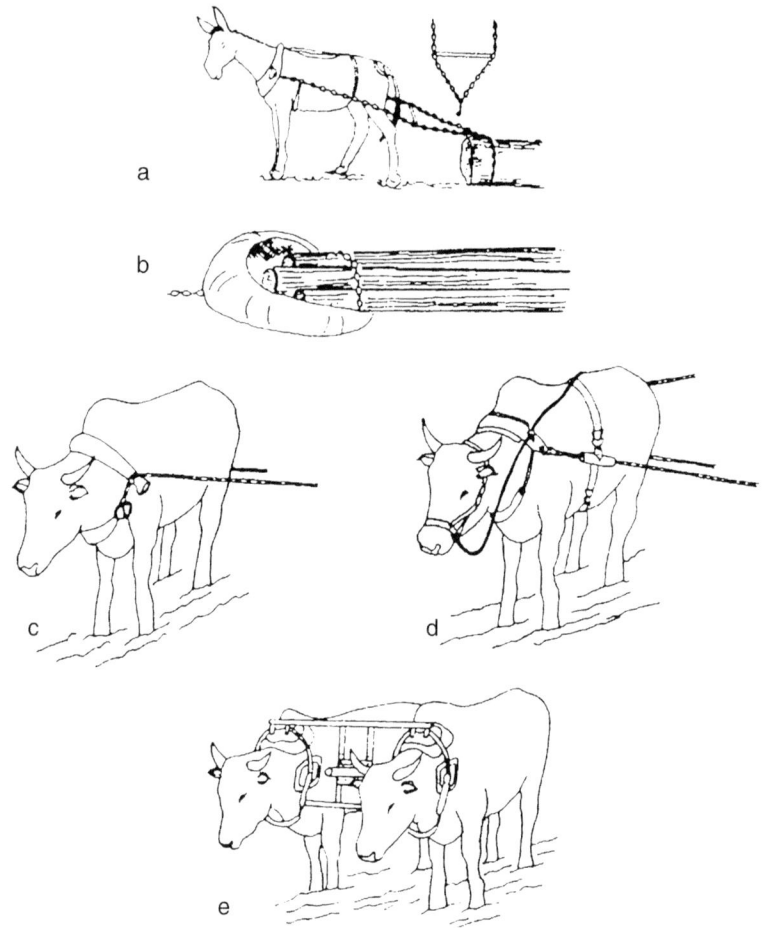

Fig. 5.1. Animal skidding. **a** Mule with harness, chain chokers. **b** Pan sledging (sledge). **c** Simple harness for a single ox. **d** Improved harness for a single ox. **e** Double-ox harness

to allow the load to drag more heavily and so provide more braking effect and protect the mule's legs from injury.

Because both mules and drivers have to work at a relaxed tempo due to the hot climate, with frequent brief rests, a full day can normally be worked. Production per day depends on skidding distance and grade and a little on volume per hectare or on log size provided the latter is such that appropriate loads can be made up.

On the basis of an average round-trip speed of 3 km/h (50 m/min) and average terminal times of 4 min per trip to hook up, unhook, and pile a load of

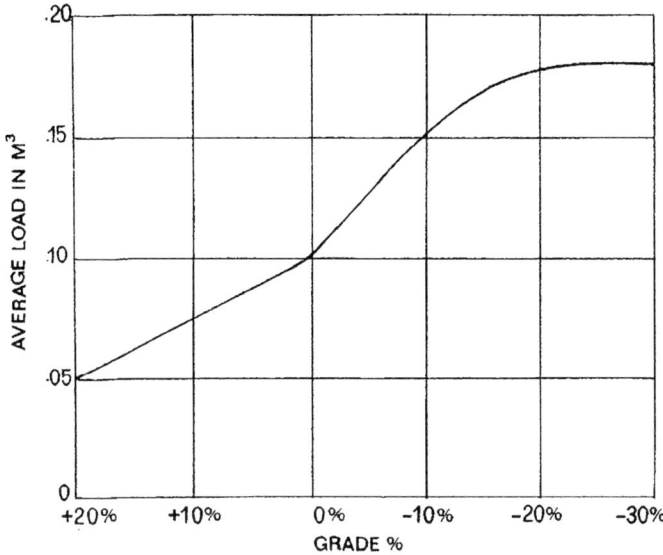

Fig. 5.2. Average load as a function of grade for chain skidding with a single mule. (Courtesy FAO 1976)

one to three logs, production per 8-h day for various skidding distances and grades is as shown in Fig. 5.3. If the effective work day is shorter or longer, production should be adjusted accordingly. The maximum payload decreases at the rate of approximately 2.5% for each percent of upgrade when skidding is uphill and increases at double that rate when skidding is downhill.

5.2
Oxen

Oxen are usually worked in pairs and are often rested on alternate days in a hot climate. They are less popular than mules but are able to skid heavier logs over more difficult terrain. Their maintenance cost is lower and their resale value higher. Maximum skidding distance is about 200 m. Oxen should not be worked on adverse grades greater than 20% or on downgrades greater than 35%. These maximum grades are marginal for efficient animal performance and should be avoided if possible. On soft soils, a maximum downhill gradient of 25% is preferred for safety reasons. On upgrades, the maximum load, as with other animals and machines, decreases at the rate of approximately 2.5% for each 1% increase in grade; thus, if a pair of oxen can chain-skid a load of 0.25 m^3 on level ground, a load of 0.19 m^3 may be expected on a 10% adverse grade

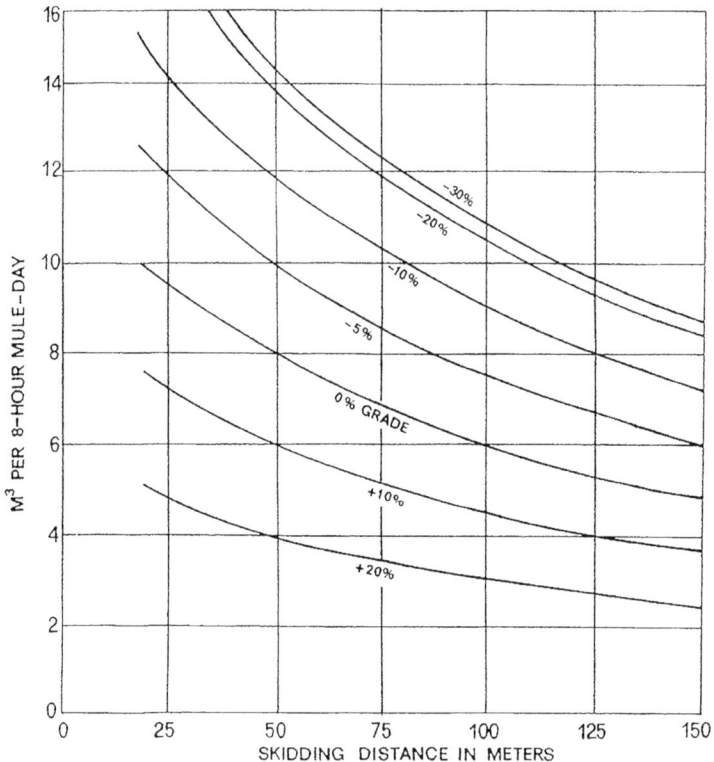

Fig. 5.3. Production per 8-h day for skidding with a single mule as a function of grade and distance. (Courtesy FAO 1976)

and 0.13 m³ on a 20% grade. On downgrades the maximum load will increase but at approximately twice the rate, i.e. about 5% for each 1% increase in grade. It will tend to level off on the steeper grades at about 0.6 m³.

The estimated maximum average loads which may be handled when chain skidding with a pair of oxen on various grades are shown in Fig. 5.4.

When a sledge is used to raise the front end of the load off the ground, the size of the load may be substantially increased (Fig. 5.4). The production per 8-h day which may be expected when chain skidding on various grades and for various skidding distances is shown in Fig. 5.5. It is based on the average travel speed of oxen of 50 m/min and an average terminal time of 7 min per trip.

When a sledge is used to skid downhill, the size of the load will tend to level off with increasing steepness because of the problems of collecting full payloads and controlling them on steep grades. The method should be used only when logs are not too heavy to be readily loaded by hand. On the basis of an average round-trip travel speed of 3 km/h (50 m/min) and average terminal

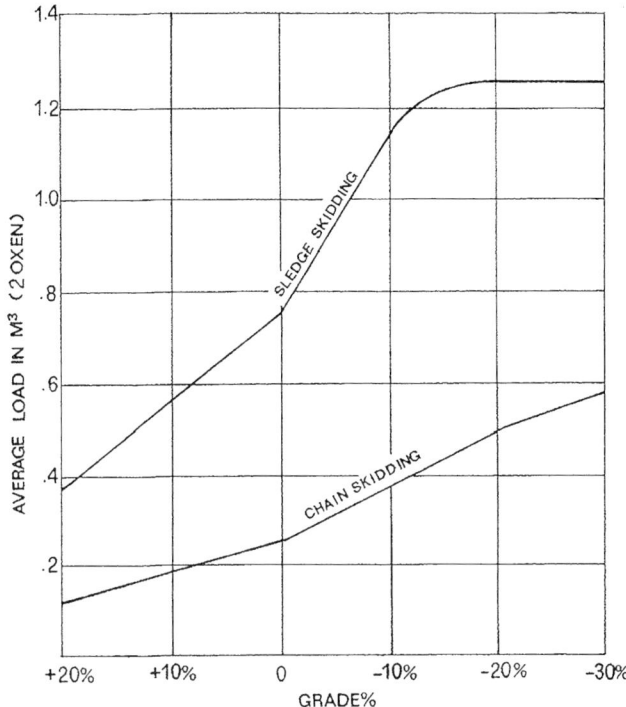

Fig. 5.4. Average load as a function of grade for chain skidding with a two-oxen team. (Courtesy FAO 1976)

times varying according to load size (from 8.5 min for a load of about 0.4 m³ to 17.5 min for a load of 1.25 m³), production per 8-h day for various grades and skidding distances is shown in Fig. 5.6. Figures should be adjusted accordingly if the length of the effective work day differs from 8 h. Although oxen travel more slowly than mules, their overall skidding speed seems to be about the same.

5.3
Elephants

The elephant is used mostly in some Southeast Asian countries, such as in Myanmar, India, Laos, Sri Lanka, and Thailand, which are situated close to the equator. Reportedly, at one time, there were as many as 12,000 working elephants in Thailand, of which 40–50% were in forest work. The weight of a full grown Asian elephant is 3–4 t and that of an African elephant 5–6 t. The African elephant has not been successfully adapted for work. Elephants eat mainly grass, as well as leaves, wild bananas, paddy rice, bamboo shoots, and

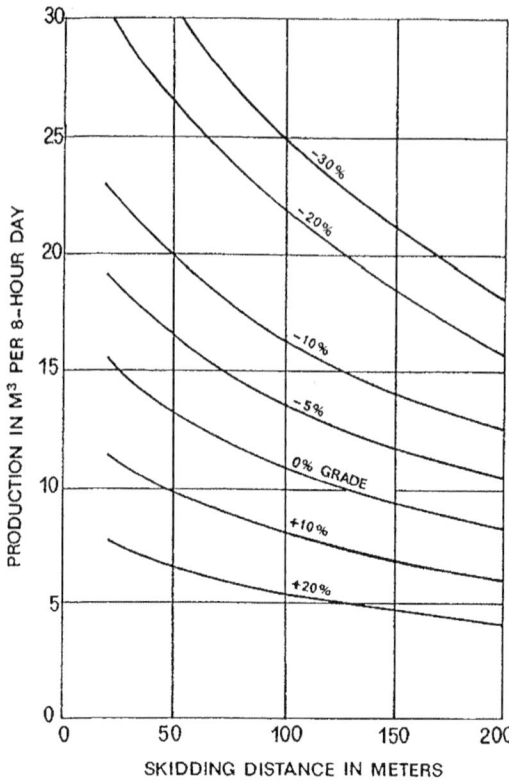

Fig. 5.5. Production per 8-h day for chain skidding with a two-oxen team as a function of grade and distance. (Courtesy FAO 1976)

twigs and bark of various trees. An elephant may consume 250 kg of grass and other foodstuffs daily, and 250 L of water; therefore, it can be used only in a few regions with a moist climate and near rivers. At the age of 3–5 years the young elephant is taken to training. A permanent rider is needed. Training lasts 5–6 years. At the age of 11–15 years the elephants may begin light work and thereafter heavy work. The peak of their working ability is between the ages of 30 and 50 years, and the retiring age is 60 years. The walking speed of an elephant is between 4 and 6 km/h. Because of this, moving them from one workplace to another takes a long time. They may be moved from place to place more quickly by truck. When elephants are used for dragging, they must be given a short rest at intervals of 500 m. When large logs are loaded with elephants, chains can be used to protect the trunk and tusks. On the basis of an average hauling distance of 1 km, an elephant can haul 450–600 m³ of timber per year in easy terrain. The capacity for work in medium or rough terrain is 300–450 m³ per year, respectively.

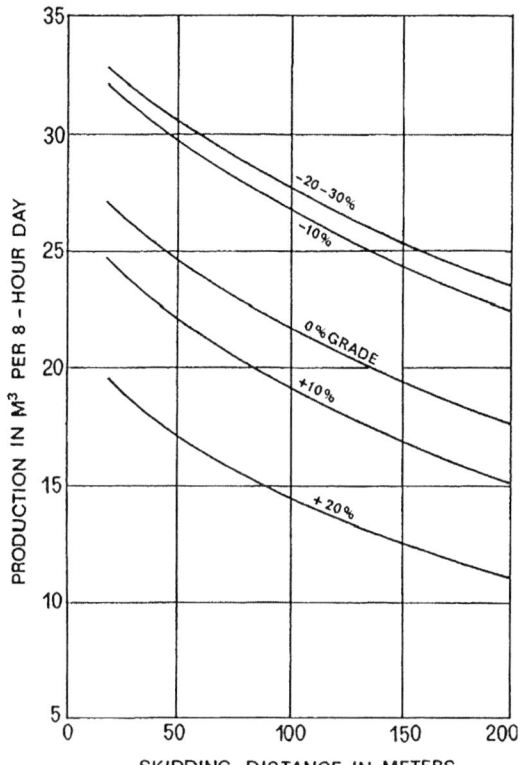

Fig. 5.6. Production per 8-h day for skidding using a sledge with a two-oxen team as a function of grade and distance. (Courtesy FAO 1976)

A single elephant can pull a log with a maximum weight of 2 t using chains of the harness attached to the log. A large log, with a weight of more than 2 t, is pulled by two elephants one behind the other. A tusker (male) is not allowed to be the rear animal of the pair, because of a danger of injury to the leading animal. Small poles reduce friction under a large log.

Elephants can roll logs with a maximum weight of 700 kg from the river bank down to the stream. A female elephant uses her trunk; a male elephant uses both trunk and tusks. With the elephant using its trunk, logs can be pushed down a hill and arranged at the landing.

A male elephant can lift and carry logs with a maximum weight of 700 kg, with its trunk and tusks, sometimes with the help of a chain; thus, it can be used for piling or loading logs. In the hot season, lifting should be avoided because of a danger of breaking the tusk.

Ground-Based Mechanized Skidding and Forwarding

6.1
Skidding Equipment

6.1.1
Natural Forest

Skidding equipment in most natural tropical forest is done by articulated rubber-tired skidders or by tracked skidders. Skidders used in natural forest are usually equipped with an integral arch, winch, and chokers (Fig. 6.1).

The rubber-tired skidder used in natural forest is normally an articulated, four-wheel-drive vehicle weighing 10–15 t with engine power of 110–140 kW. It is equipped with a blade for light pushing of obstacles and stacking of logs. It is fast, with a maximum speed of 25–30 km/h.

The tracked skidders can be divided into rigid and flexible tracks depending upon the type of suspension system that is used. Rigid-track skidders, also referred to as crawler tractors, are the most common type of skidding machine used in natural forest. Flexible-track skidders permit the track roller (road wheels) to move independently, while rigid-track skidders do not. The crawler tractor used for road construction normally has a rigid track. Rigid-track skidders are strong, slow machines built for pushing and heavy pulling under adverse conditions. Rigid-track skidders normally used in natural forest vary from engines of 120 to 160 kW and have a maximum speed of about 12 km/h. Flexible-track skidders are built for high speed under weak ground conditions and some can travel almost as fast as rubber-tired skidders. Although they have a blade, it is used for light work such as piling logs or removing debris in a similar way as for rubber-tired skidders. The flexible-track construction is not constructed for the heavy turning work typical of construction work. Tracks can come off flexible-track skidders more easily than with rigid-track skidders.

The rubber-tired skidder is faster and less expensive than the tracked-skidder, but it can develop less traction and cannot pull as large a load as the tracked skidder. Thus, the rubber-tired skidder is normally used where the

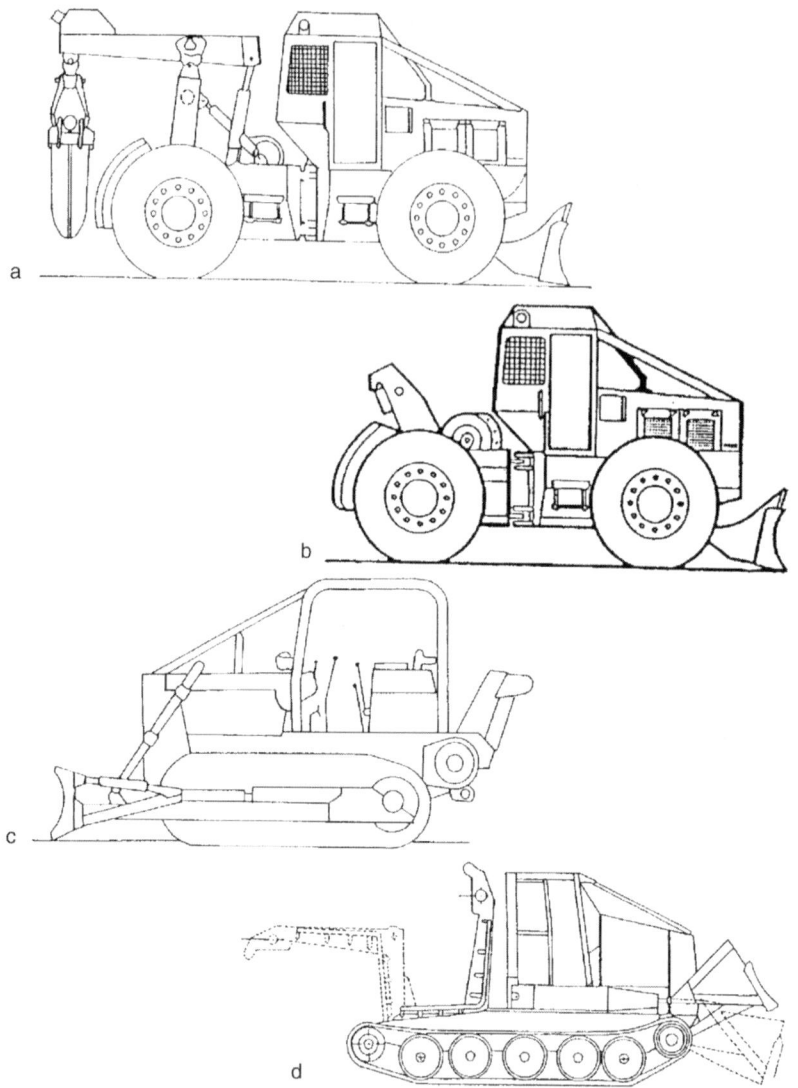

Fig. 6.1. a Rubber-tired skidder with a grapple. **b** Rubber-tired skidder with a winch. **c** Rigid-track skidder with integral arch and winch. **d** Flexible track skidder with a choker arch

ground conditions are better, log size is not excessive, and the skidding distances are longer. Rigid-track skidders are used where logs are large, skidding distance is short, and ground conditions are poor. Flexible-track skidders are used where skidding distances are long, traction is poor, but ground conditions are not rough. All skidders work best when skidding downhill. The maximum safe slope for rubber-tired skidders to operate on is about 30%. Tracked skidders can operate on slopes up to about 40%. In tropical high forest, rubber-tired skidders are exclusively used on skid roads and trails.

Logs are usually choked with wire rope that is then attached to the skidding line. The skidding line is attached to a powered winch on the skidder. In order to attach the choked logs to the skidding line, the skidding line is pulled from the skidder to the logs. The force required to pull the skidding line to the logs is a function of the weight of the line, the distance to the logs, and the slope. Pulling wire rope skidding line is heavy work, especially in the tropics. Recently, synthetic rope has been introduced as an alternative to wire rope. Synthetic rope has almost the same strength as wire rope but weighs only a tenth as much. This reduces the effort for pulling the skidding line. Synthetic rope is about 3 times as expensive as wire rope.

Sometimes, crawler tractors are used to break logs out of their bed and to skid them to a main skid trail of higher standard where either a rubber-tired skidder or a flexible-track skidder is used to skid (swing) the logs a longer distance to a truck road. Skidders used for swing operations are sometimes equipped a grapple to eliminate the rehooking of the logs.

6.1.2
Planted Forest

Skidding in planted forest is normally done by agricultural tractor, rubber-tired skidder, or forwarder (Fig. 6.2). The agricultural tractors are either two-wheel drive or four-wheel drive. The skidders can be articulated or nonarticulated, two-wheel drive or four-wheel drive. Some are three-wheel with a two-wheel hydrostatic drive. The tractors and skidders might use either chokers or grapples. Power and weight depend upon tree size.

Forwarders are articulated rubber-tired vehicles with a log bunk to carry logs free of the ground. Forwarders usually carry 2.5–6-m logs in loads of 5–15 t and have engines of 70–150 kW. Swingboom grapples to load and unload the log bunk are used to carry the load. Advantages of forwarders over skidders include delivery of cleaner wood, lower travel resistance due to elimination of dragging logs on the ground, and not having men hooking or unhooking logs on the forest floor. However, forwarders are more expensive than rubber-tired skidders and cannot deliver as long a log.

A compromise between the skidder and the forwarder is the clambunk skidder (Fig. 6.2). The clambunk skidder has an inverted grapple behind the swingboom grapple which is used to hold long logs or trees. No men are on the forest floor and the load is unloaded by opening the inverted grapple and driving out from underneath the load. Clambunk grapple skidders can be

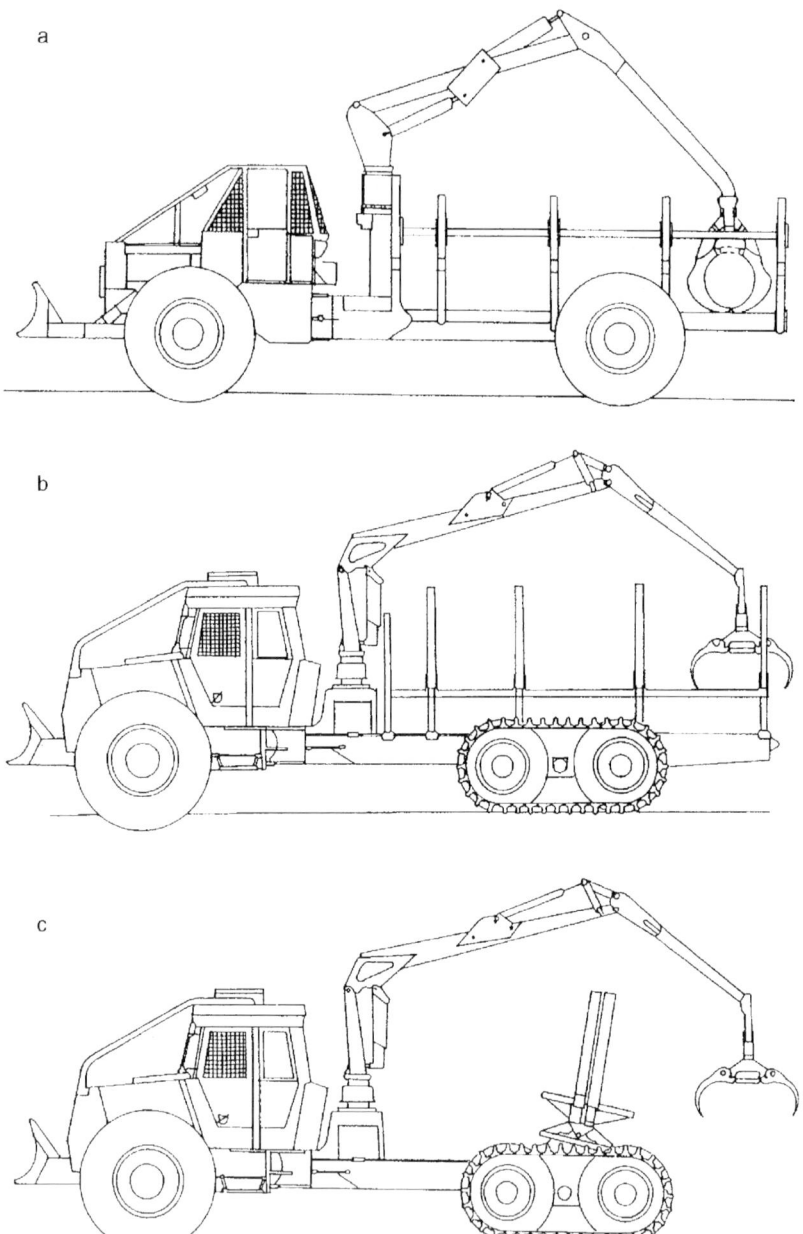

Fig. 6.2. **a** Forwarder for 5.0-m wood. **b** Forwarder (with bogie and steel tracks) for 7.5-m wood. **c** Clambunk skidder

either rubber-tired or tracked. Clambunk skidders are not common in tropical manmade forest.

Although not common in planted forests, in tropical forests modified hydraulic log loaders can be used to swing either tree lengths or log lengths from stump to roadside. Termed shovel logging, because hydraulic log loaders evolved from hydraulic excavators, these machines have proved economical in western and southeastern North America for final harvest in even-aged management. Using this system, one operator and one machine can both swing the logs from stump to roadside as well as load trucks. Shovel logging machines generally have longer, wider tracks, higher clearance, and heavier track drives than on-road log loaders (Fig. 6.3). The hydraulic cylinders are located under the boom similar to a hydraulic log loader as opposed to an excavator.

6.2
Skidding and Forwarding Patterns

6.2.1
Natural Forest

The planning of skidding and felling is important. Preplanning of skid trails and directional felling will improve skidding efficiency, increase safety, and reduce ground disturbance. Depending upon the terrain, road and landing construction costs, and volume per hectare to be removed, some type of uniform system can usually be organized (Fig. 6.4). On steeper terrain, either a branching or a parallel pattern can be used (Fig. 6.5).

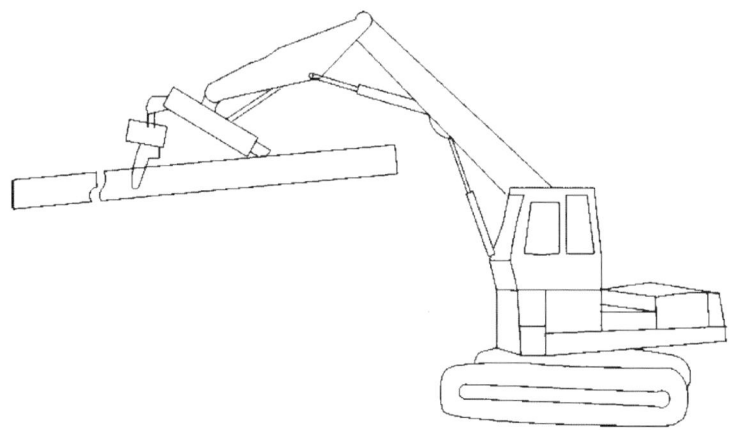

Fig. 6.3. Shovel logging machine

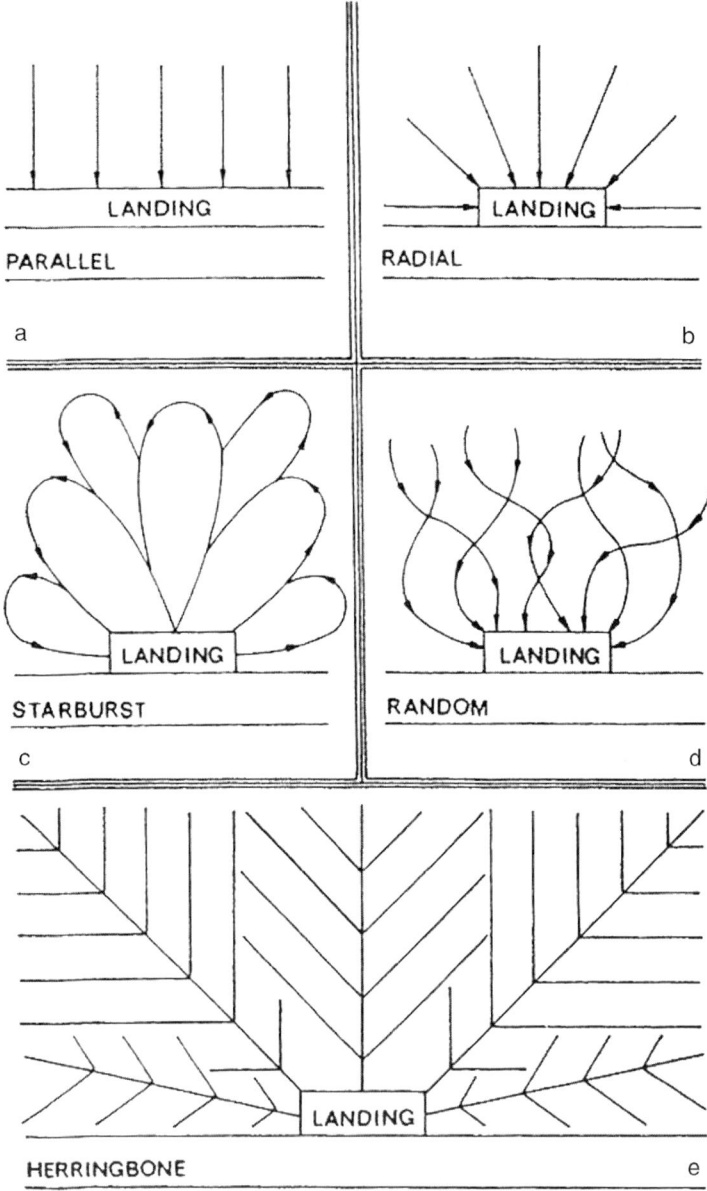

Fig. 6.4. Skidding patterns. **a** Parallel. **b** Radial. **c** Starburst. **d** Random. **e** Herringbone

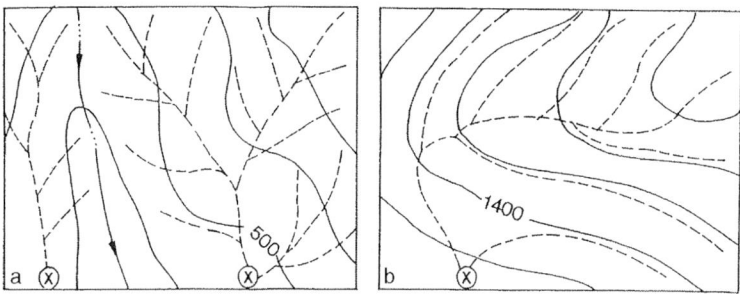

Fig. 6.5. Skidding patterns on steeper terrain. **a** Branching. **b** Parallel

In natural forest with trees with large crowns, a combined felling and skidding operation (hot-logging) works well:

- Spur roads and landings are located using the optimal road spacing as a guide.
- Landing size is usually about 0.1 ha. Landings are located at least 30 m from environmentally sensitive areas on well-drained or easy-to-drain locations.
- Major skid trails are flagged. If rubber-tired skidders are to be used, a pass with a crawler tractor is used if necessary to clear the skid trails.
- A supervisor assigns each crew to a skid trail. A crew consists of one to two cutters, and a choker setter.
- Cutting begins at the end of the skid trail falling the trees away from the skid trail so that the skidders do not have to skid through the large hardwood tree crowns.

6.2.2
Planted Forest

6.2.2.1
Skidders

Skid trails are laid out to facilitate the silvicultural plan for the stand. A more or less parallel system of skid trails can be determined using the optimal road spacing as a guide. Often it will be in the range of 30–60 m (Fig. 6.6). Trees are felled in a herringbone pattern toward the skid trails to facilitate skidding and minimize damage to the residual stand. Skidders work only on these skid trails to prevent damage to the soil and trees. Skidding of the wood to the trails is done by cable winch, animals, or manually depending upon tree size, equipment and labor costs, and social policy. Sometimes, at the time of clear

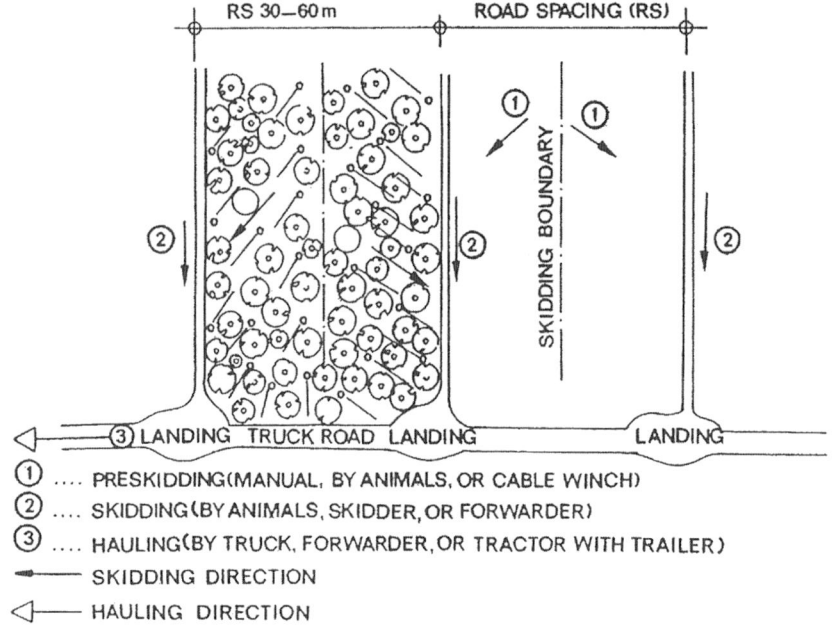

Fig. 6.6. Felling pattern and skid trail spacing

felling, the parallel skid trails will be replaced by a radial set of skid trails to the landing if soil compaction is not a problem.

Skid trails for mechanized felling and bunching are also parallel, but are more closely spaced, 20–25 m (Fig. 6.7). The operations can be done in several ways. One way is for the feller-buncher to initially start cutting a 3–4-m-wide strip, laying all bunches to the right. As soon as the end of the strip is reached, the operator goes to his left and begins selectively thinning the between-strip space up to where the next the strip will be.

The trees from the selective area would be bunched in the previously cut strip, thereby minimizing damage and also concentrating all of the cut material either in the cut strip or adjacent to it. After the operator has finished the selective cutting, he is back at his starting point and has one cut strip and a 20–25-m band of selectively thinned area. He then moves 20–25 m to the left and begins a new strip, again laying bunches to the right and into the thinned area he just finished cutting. As he progresses in this manner to his left, he can cut many hectares and never interfere with the skidder. With all the material being concentrated on a strip, the skidder operator, experienced or not, can easily follow the cutting pattern without having to make any skidding route decisions.

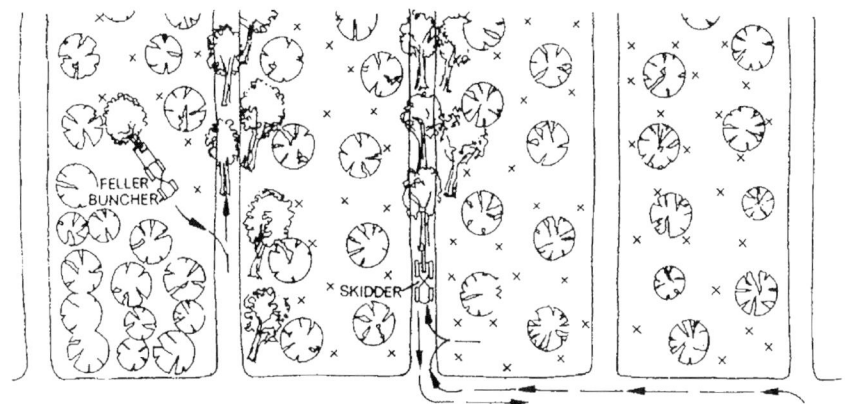

Fig. 6.7. Felling with feller-buncher and skidding with a grapple skidder

If a grapple skidder is used, the feller-buncher operator must be familiar with the maximum size load the grapple skidder can handle in order to optimize the latter's productivity. Too narrow a strip prevents the grapple skidder from attaining maximum productivity.

6.2.2.2
Forwarders

Forwarder trails (strip roads) are also laid out in a parallel manner. For thinning, the forwarder trails are farther apart (20–30 m) than at final harvest (15–18 m), but must be within the reach of the grapple. The forwarder trails should be as straight as possible. Steep change of grade, tight bends, or junctions should be avoided. Trails should run with the slope, not the sidehill to prevent overturning and should be long enough to collect at least one full load. Most forwarders have reversible controls so the operator can drive "backwards" empty and "forward" loaded. On long trails intermediate connecting roads can shorten return time. Wood for forwarders is usually cut in approximately 2.5-m lengths (Fig. 6.8) or 5.0-m lengths (Fig. 6.9) depending upon how it will be transported to the mill.

Sometimes trees in the forwarder trails are felled first parallel to the trail and bunched. Then remaining trees are felled toward trails and bunched. Slightly faster loading times are achieved if the operator does not have to rotate the grapple each cycle or if the rotation is minimized. Depending upon soil type, rainfall intensity, trail gradient, tree species, and harvesting method, the slash in the strip road is sometimes left or placed in the trails to increase bearing capacity and reduce erosion.

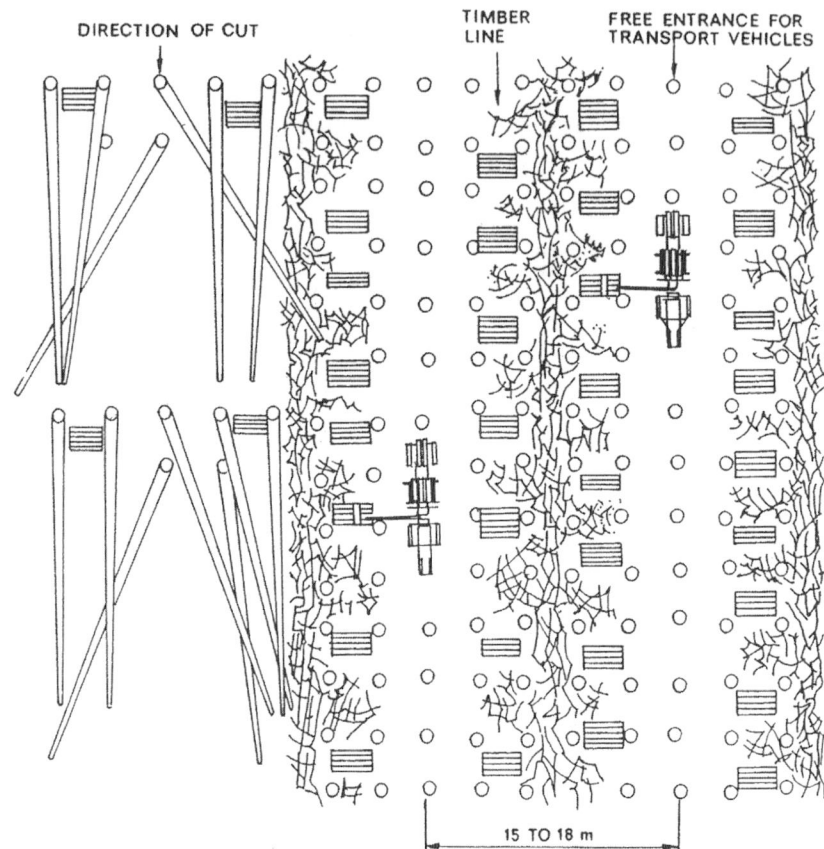

Fig. 6.8. Forwarder trail piling and spacing for 2.5-m shortwood

6.2.2.3
Shovels

Several shovel logging patterns are used, but all more closely resemble continuous landings as opposed to discrete landings. There are two common shovel logging patterns. The first is a serpentine pattern (Fig. 6.10) where the shovel begins at the back of the unit and works its way forward accumulating the wood in rows, or racks, and moving the rows, or racks, forward. Sometimes an initial pass is made along the road to straighten the logs that will ultimately form the base of the log deck. The serpentine pattern is most commonly used on flat terrain with long logs. An alternative pattern, more often used in sloping terrain (up to 40% slope) or with full trees, is to travel on trails perpendicular to the road (Fig. 6.11). Shovel logging cost increases with distance.

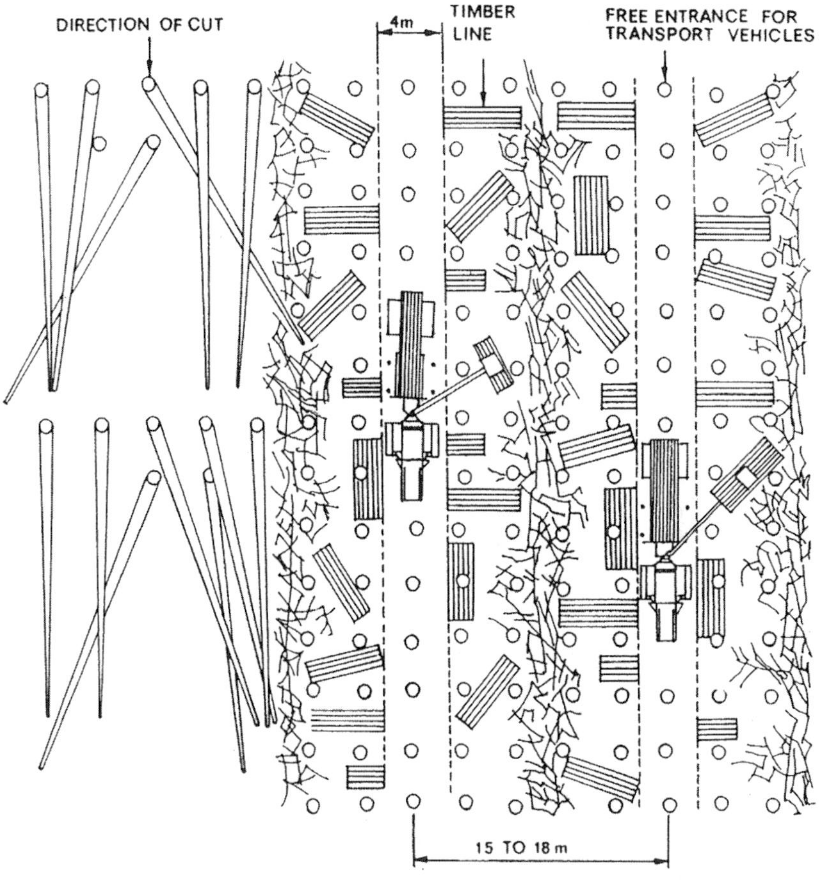

DIRECTION OF CUT
4m
TIMBER LINE
FREE ENTRANCE FOR TRANSPORT VEHICLES
15 TO 18 m

Fig. 6.9. Forwarder trail piling and spacing for 5.0-m wood

1
2
3
4
5
Start

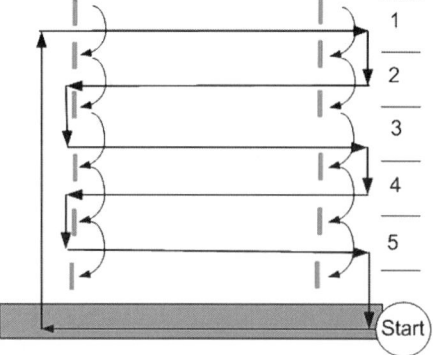

Fig. 6.10. Serpentine shovel logging pattern for swinging long logs on gentle terrain. The operator often starts at the lower right straightening up the rack closest to the road, then proceeds to the back of the unit, and works toward the front of the unit in a serpentine pattern

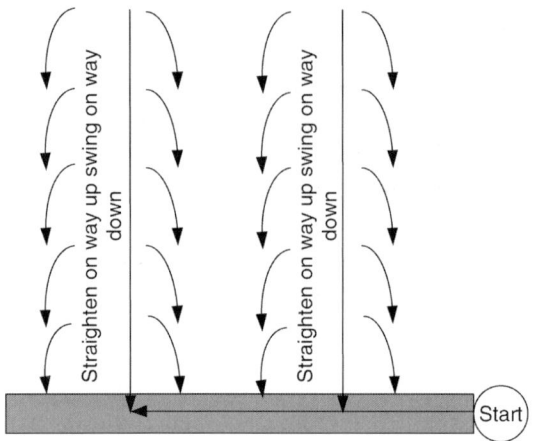

Fig. 6.11. Vertical pattern for swinging tree length or full trees to the roadside and for operations on steeper terrain. The shovel arranges trees or logs while preparing a trail perpendicular to the truck road and then returns along the trail swinging the trees or logs to the roadside

6.3
Production and Cost

To estimate skidding production in cubic meters per productive machine hour (PMH) and skidding cost per cubic meter (SCM) the following formulas can be used:

$$PSM = \frac{60L}{TT + \dfrac{ASD}{ATSL} + \dfrac{ASD}{ATSE}},$$

and

$$SCM = \frac{C + c(1 + f)}{PSM},$$

where PSM is the production in cubic meters per PMH, L is the skidder load in cubic meters, TT is the fixed time per load in minutes (loading, unhooking, and delay per trip), ASD is the average skidding distance in meters, ATSL is the average travel speed loaded in meters per minute, ATSE is the average travel speed with the machine empty in meters per minute, c is the direct wages of the operator per PMH, f is the cost of fringe benefits expressed as a percentage of direct wages, and C is the operating cost of the skidder per PMH, which can be estimated using the machine cost formula.

It is possible to calculate the theoretical travel speed for skidders and forwarders based upon the equipment specifications, load size, and terrain conditions. However, conditions can be quite variable and the best guide is often local experience. Measure the vehicle speed and load size under the terrain conditions expected during skidding operations. This can provide a rapid estimate of machine performance.

Figures 6.12 and 6.13 give some ideas of production per effective working day (shift) during the favorable season in natural forest. The long-run production will not always correspond to figures derived from the graphs. "Favorable season" refers to conditions during the dry period in tropical climates. The long-run production will be reduced by low output or shutdown during wet periods. Experience shows that during wet periods the production will often drop to 65–75% of the corresponding production during dry periods.

It is important to make realistic assessments of the length of the working period in different seasons. Besides the climate, machine maintenance and service facilities are significant aspects in this respect. Experience shows that only 50–70% of the available number of working days (some 250 annually) are effectively used for skidding (because of breakdowns, waiting, bad

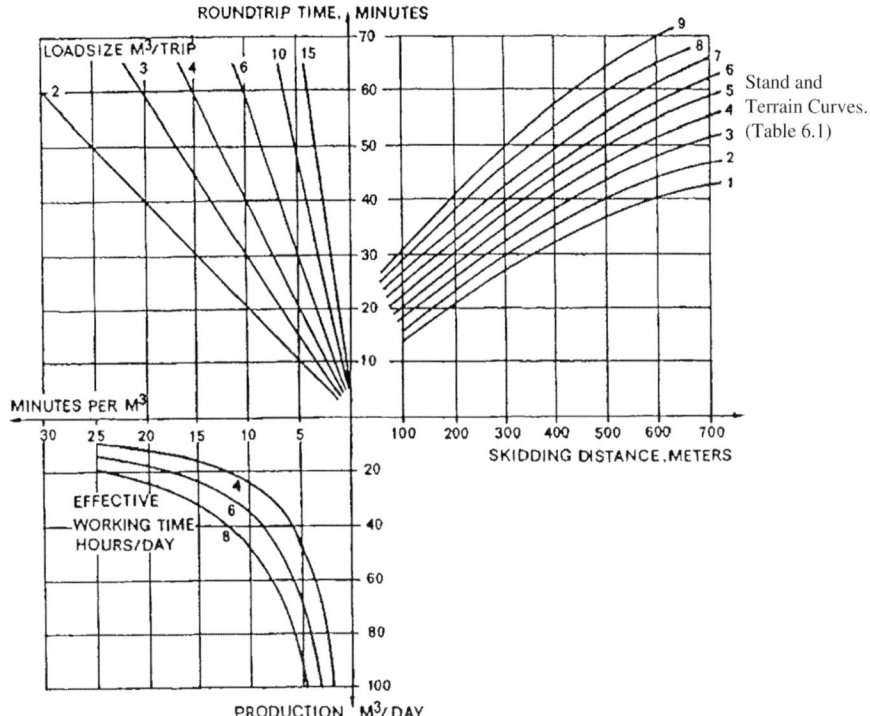

Fig. 6.12. Skidding production in natural forest with a rigid-track skidder. See Table 6.1. (Courtesy FAO 1976)

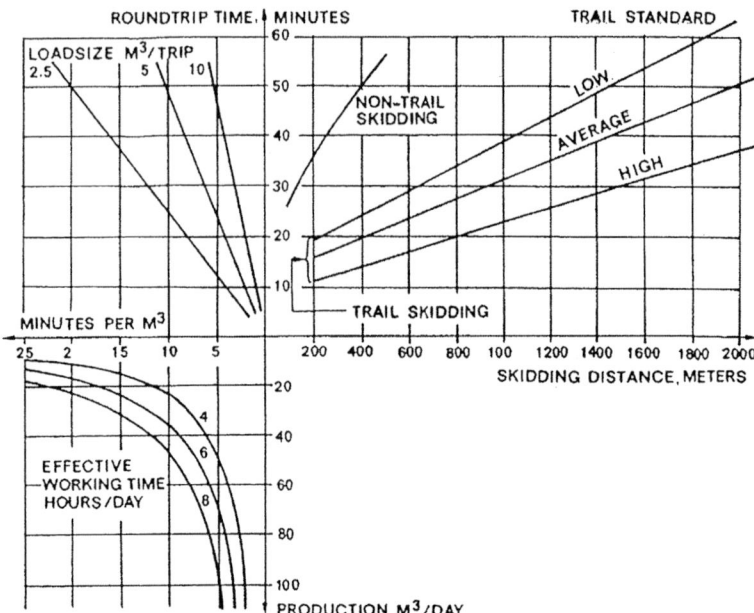

Fig. 6.13. Skidding production in natural forest with a rubber-tired skidder. (Courtesy FAO 1976)

weather, other tasks for the equipment, etc.). Despite a high daily production, the annual output of a large crawler tractor is often found to be not more than 4,000–5,000 m³ (although the theoretical production might be 3–4 times higher). The output that can be achieved depends strongly on load size; however, a small load size is sometimes compensated by reduced round-trip time.

Figure 6.14 shows the typical relationship between load size and net power of the skidding machine. The graph seems to be reasonably valid for both crawler tractors and articulated wheeled skidders. However, exceptions from this observation are not uncommon. In large timber, crawlers usually skid the biggest logs, whereas skidders skid the small and medium-sized logs. Also, the range of load sizes seems to be wider for crawlers.

Figure 6.14 is based on average load sizes as recorded in available sources. The diagram is very simplified since the load size is sensitive to different conditions in the work environment (harvested volume per hectare, density of undergrowth, terrain, and species density). The load size depends also on the type of forest harvested and on the logging system design.

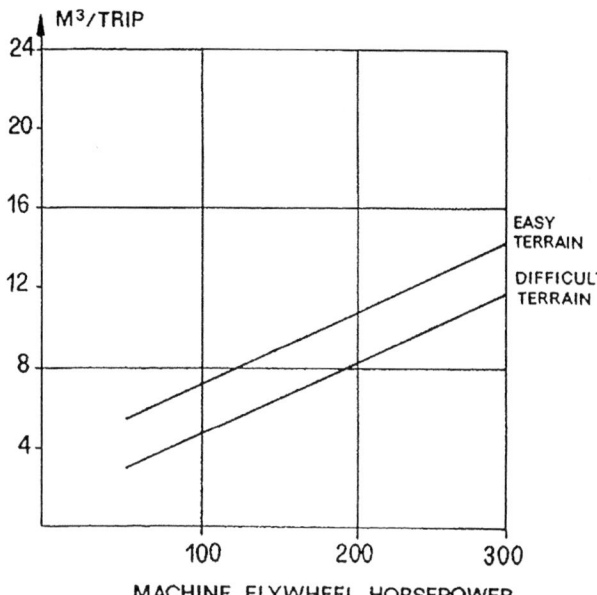

Fig. 6.14. Typical load sizes for tracked and rubber-tired skidders in natural forest. (Courtesy FAO 1974)

The following procedures are suggested for the assessment of load size for two situations:

1. Skidding machines already on hand
   (a) Read load size over net horsepower rating of the machine in Fig. 6.14.
   (b) Check that this load size is likely to be achieved with regard to terrain; skidding uphill on grades of more than 10–15% will reduce the practical load size; skidding downhill on grades of more than 35–40% will reduce practical load size.
   (c) Check if soil conditions under "normal weather conditions" will permit the load size.
   (d) Check if tree and log size will permit a full load; in tropical high forest skidding, one log per load is normal and more than two logs per load is unusual.
   (e) Check if harvested volume per hectare permits the estimated load size.
2. Choice of skidding machine to be made and roads to be constructed
   (a) Estimate average volume of merchantable bole and maximum log size, taking into account that the biggest boles may be cut into two or several logs.

(b) Decide whether ridge roads or valley roads will be constructed and thus if uphill or downhill skidding will dominate; enter load size in Fig. 6.14 and read the corresponding net horsepower rating on the horizontal axis, considering the aggregate influence of terrain, soil, and climate as shown by the lines called "easy terrain" and "difficult terrain."

(c) Choose a skidding machine corresponding to the net horsepower rating.

Example of estimating skidding production: Estimate the daily production of a 175-hp rubber-tired skidder in difficult terrain working six effective hours per day. The trail standard is average and the average skidding distance is 800 m.

- Step 1. Find average load size from Fig. 6.14: 7 m^3 per trip.
- Step 2. Find production per 6-h effective day from Fig. 6.13: 75 m^3.

6.3.1
Rigid-Track Crawler Tractor

Production forecasts for rigid-track crawler-tractor skidding are based on Fig. 6.12. It is more appropriate to use Fig. 6.13 for flexible-track skidders owing to their greater speed. Starting with the skidding distance (the horizontal axis in the nomograph) one of the nine numbered curves is selected in order to arrive at the expected round-trip time (vertical axis). The curve selected is determined by Table 6.1, which gives a curve number after evaluating slope, harvested volume per hectare, type of soil, and obstacles.

Table 6.1. Table for selecting stand and terrain curve used for estimating rigid-track skidder production to be used with Fig. 6.12

	Slope (% downgrade loaded)								
	0–15			15–35			Over 35		
Harvested volume (m^3/ha)	5	25	100	5	25	100	5	25	100
Firm, dry soil; little downtimber, rocks, or undergrowth	3	2	1	4	3	2	8	6	5
Moderately wet or soft soil; some downtimber, rocks, or undergrowth	4	3	2	5	4	3	8	7	6
Muddy or loose soil; much downtimber, rocks, or undergrowth	6	4	3	7	6	5	9	8	7

The production estimate is completed by establishing in Fig. 6.12 the load size per trip (with the help of Fig. 6.14 if no local information is available) and the effective working hours per day. The nomograph includes all phases of work: loading, unloading, skidding, and return trip. There are, of course, very complicated relationships between time consumption in these different phases and load size, type of tractors, terrain difficulty, log size, etc. The charts and table are designed to cut through all such interrelations.

If the skidding route varies regarding slope, soil conditions, or any other characteristic, a mean value may be established. Should the variations be very significant, it is advisable to divide the area in two or three reasonably homogeneous parts and carry out the exercise for each part separately.

6.3.2
Rubber-Tired Skidders

There are two different situations for the use of rubber-tired skidders:

1. Work takes place on prepared skidding trails. In this case the skidder generally works along with a crawler tractor, one of the tasks of the crawler being to prepare the trails. This is "trail skidding."
2. Work takes place as "nontrail skidding," which means that the skidder works on the undisturbed forest floor. In tropical high forests, rubber-tired skidders do not normally work on the undisturbed forest floor.

Production forecasts for the two situations are based on Fig. 6.13, which is read starting with the skidding distance and reading counterclockwise.

As in the case of the crawler-tractor skidding, there are complicated relationships between load size, trail standard, and other conditions. The assessment of trail standard in Fig. 6.13 should be based on the following average speeds, excluding delay times, for skidding and return trips:

Trail standard	Average speed in km/h
Low	<8
Average	8–12
High	>12

In Figs. 6.12 and 6.13 production factors are skidding distance, load size, and the number of effective hours per day. Work environment such as terrain and slopes is included to a certain extent.

The influence of the above conditions on output is obvious. They determine either the number of round trips per day or the production of each round trip.

Terrain conditions influence production in two ways. Primarily, the terrain influences the travel speed of the machine and thus the round-trip time. Secondly, the load size has a direct but limited relationship to the terrain (Fig. 6.14). Unfavorable terrain conditions tend to result in smaller loads.

6.3.3
Flexible-Track Skidders

The flexible-track skidder has performance characteristics which combine some of the advantages of the crawler tractor and the rubber-tired skidder. Its traction capability is as good or better than that of the rigid-track crawler tractor and its speed is comparable to that of the rubber-tired skidder. However, it is more expensive than a rubber-tired skidder and its track system requires more operating care and maintenance than a rigid-track crawler tractor. It is more appropriate to use Fig. 6.13 for estimating production for flexible-track skidders owing to their greater speed than to use the nomograph for rigid-track crawler tractors (Fig. 6.12).

6.3.4
Combined Use of Crawler and Wheeled Skidders

There are basic differences between the rigid-track skidder and the rubber-tired skidder that make them suited for different phases of work. The crawler is expensive, powerful, and travels with limited speed, whereas the wheeled skidder is cheaper, travels faster, but has less traction than the crawler. In crawler skidding the tractor can push its way to the felled tree through the remaining stand. It constructs a simple road between the landing and the tree. However, closer to the landing, the tractor can operate on trails prepared during previous trips. For this part of the route the crawler's penetration capability is no longer needed and this phase of the operation is taken over by the wheeled skidder. The result is that the wheeled skidder operates on prepared trails transporting logs to the landing and the crawler operates in the upper part of the transport system bringing the logs from stump site to the trail. With this organization the crawler can make use of its power and the wheeled skidder of its speed. The crawler tractor is used generally as the primary mover, handling the more difficult part of the skidding. Figure 6.12 reflects this operation if curves with high numbers are used. Generally two crawler tractors are used in combination with one rubber-tired skidder. In more difficult terrain, the combinations could be even 3:1. The use of the crawler tractor and wheeled skidder in combination will mean lower costs for skidding and/or less truck road. Calculations indicate that when the two machines are used together as described earlier, the optimum distance between truck roads will increase by 20–30%, i.e., the optimum length of truck road (meters per hectare) in the logging area will be reduced by 15–25%.

6.3.5
Forwarders

An estimate of forwarder production and cost can be derived with the following method. Determine:

- Stand density in cubic meters per hectare
- Log length in meters
- Forwarder payload capacity in cubic meters
- Grapple closed area or diameter (Table 6.2)
- Forwarder operating cost per hour
- Average strip width in meters

Table 6.2. Calculated capacity in cubic meters of grapples of various sizes

Closed grapple		Log length (m)					
		1	2	3	4	5	6
Area (m²)	Inside diameter (cm)	F=0.67	F=0.64	F=0.61	F=0.58	F=0.55	F=0.52
0.20	50	0.14	0.26	0.37	0.47	0.55	0.63
0.25	56	0.17	0.32	0.46	0.58	0.69	0.78
0.30	62	0.20	0.38	0.55	0.70	0.83	0.94
0.35	67	0.23	0.45	0.64	0.81	0.96	1.09
0.40	71	0.27	0.51	0.73	0.93	1.10	1.25
0.45	76	0.30	0.58	0.82	1.04	1.24	1.40
0.50	80	0.34	0.64	0.92	1.16	1.37	1.56
0.55	84	0.37	0.70	1.01	1.28	1.51	1.71
0.60	87	0.40	0.77	1.10	1.39	1.65	1.87

The grapple capacity is the product of the area in square meters, the log length in meters, and F, where F is the ratio of the wood volume inside the bark to the stacked volume of unbarked wood. One cubic meter is 1.67-m³ stacked volume. The value of the F will vary somewhat with log length because longer logs tend to lie less closely together when grouped.

To obtain production in cubic meters per PMH:

- Read the grapple capacity in Table 6.2.
- Knowing stand density and strip width, read the strip length per cubic meter of forwarder load in meters (Table 6.3).
- Take average grapple loads to be 0.70 times the grapple capacity when loading and 0.90 times the grapple capacity when offloading.
- Take the average grapple cycle time to be 0.50 min or field observation.
- Take the average forwarder travel speeds on level clean ground to be 40 m/min while loading in the strip and 60 m/min on the trail to and from the roadside.
- Find the average forwarding distance.
- Estimate the value of the time adjustment factors (TA) that should be applied to compensate for poor terrain conditions, climate, operator training, operator skill and motivation, and personal delays.
- Apply the formula.

$$
\text{PFM} = \frac{60L}{\dfrac{0.50L}{0.70GC}(1+\text{TAP}) + \dfrac{L \times \text{SL}}{\text{ATSS}(1-\text{TAT})} + \dfrac{2\text{AFD}}{\text{ATS}(1-\text{TAT})} + \dfrac{0.50L}{0.90GC}(1+\text{TAP})},
$$

where PFM is the production in cubic meters per PMH, L is the forwarder payload in cubic meters, GC is the grapple capacity in cubic meters, SL is the strip length per cubic meter of payload, ATSS is 40 m/min (the average travel speed while loading in the strip under optimum conditions) or use field observation,

Table 6.3. Strip length per cubic meter of forwarder payload

Volume per hectare (m³)	Strip width (m)		
	10	15	20
20	50	33	25
30	33	22	17
40	25	17	12
50	20	13	10
60	17	11	8
70	14	10	7
80	12	8	6
90	11	7.5	5.5
100	10	6.5	5
120	8	5.5	4
150	6.5	4.5	3.5
200	6	3.5	2.5

AFD is the average forwarding distance in meters, ATS is 60 m/min (the average travel speed while traveling to and from the roadside) or use field observation, TA is the time adjustment factors to be applied to basic productive time, TAT is the terrain adjustment factors, and TAP is the personal adjustment factors.

The forwarding cost per cubic meters is then found with the formula

$$\text{FCM} = \frac{C + c(1+f)}{\text{PFM}},$$

where FCM is the forwarding cost in dollars per cubic /meter, c is the direct wages of the operator per PMH, f is the cost of fringe benefits expressed as a percentage of direct wages, PFM is the forwarding production in cubic meters per PMH, and C is the cost per PMH.

Example for estimating forwarding production and cost: Estimate the forwarding production and cost for a 14-m^3 capacity forwarder with 0.5-m^2 grapple capacity operating on forwarder trails 15 m apart removing 150 m^3/ha in stacked 5-m logs. The forwarder is traveling with a speed of 40 m/min during loading and with a speed of 60 m/min when traveling to and from the loading area. The average forwarding distance is 1,000 m. There is a 10% increase for terrain conditions, and a 15% increase for operator skill and personal delays.

- Step 1. Find the grapple capacity from Table 6.2: 1.37 m^3.
- Step 2. Find the strip length per cubic meter from Table 6.3: 4.5 m.

$$\text{PFM} = \frac{60 \times 14}{\dfrac{(0.50 \times 14) \times (1 + 0.15)}{0.70 \times 1.37} + \dfrac{(14 \times 4.5)}{40 \times (1 - 0.1)} + \dfrac{2 \times 1000}{60 \times (1 - 0.1)} + \dfrac{(0.5 \times 14) \times (1 + 0.15)}{0.90 \times 1.37}}$$

$$= 15.6 \, \text{m}^3/\text{h}.$$

If the forwarder cost is $70 per hour, the operator cost is $4 per hour and fringe benefits are 40%, then

$$\text{FC} = \frac{70 + [4 \times (1 + 0.4)]}{15.6} = \$4.85 \text{ per cubic meter}.$$

6.4
Landings for Ground-Based Systems

A landing is an open or cleared area where logs or tree lengths are brought by skidders and ultimately loaded on trucks for road transport. While the logs or tree lengths are on the landing, a number of operations may be done, including species identification, tree measurement, bucking, limbing, log scaling, log grading, log marking, sorting, or a combination of these activities, before logs are loaded on trucks for road transport. Planning for the landing is very important to the efficiency, safety, and environmental impacts of the operation. A smooth-flowing landing operation encourages productivity in skidding and road transport. A landing with some storage capacity also uncouples skidding production from truck transport. In that way, neither operation is dependent upon the other. The landing should be located in a well-drained area, with room for skidding equipment to enter, unload, and turn around, for log processing activities to safely take place, for log storage, for debris storage, and for truck loading. Landings should be located so that free drainage occurs at all times. Ideally landings should be located on gently sloping elevated areas. A minimum slope of 2–5% is recommended. Landings should be located at least 30 m away from sensitive areas.

In wet, clay soils, where rock is not available, a corduroy surface of non-merchantable small logs and branches can be placed under areas where the loader, trucks, and equipment maintenance vehicles may pass. Skidders should not be allowed to enter the road. Ideally, during the wet season, log loaders should also be kept off the road to prevent contamination of the road surface. If possible, trucks should remain on the road during loading to prevent bringing contaminants onto the road surface and to maintain road ditch integrity. Once the surface of a forest road is destroyed, subgrade failure can quickly take place and during the wet season, road restoration will be difficult and expensive.

A variety of log landing designs can be used (Figs. 6.15–6.19). Landings for ground-based equipment are often rectangular with main skid trails entering on the shorter sides. Landing size depends upon the activities that will be conducted on them and the level of activity. For example, tree-length skidding to landings will require larger landings than log-length skidding. For tree-length skidding, the landing will need to be longer and wider to provide an area for tree measurement and log bucking. Landings where log trucks cannot be loaded directly on the spur road will be larger than landings where the trucks can remain on the road. Landings where loaders are not permitted on the road will be larger than those where loaders can operate on the road. Landings that require many log sorts will need to be larger than those where no sorting is done. In many cases, log landings can be kept to approximately 0.1 ha (1,000 m^2), but often there are good reasons to exceed this size. A single log landing could serve 20–40 ha or more. If a 0.1-ha landing serves 20 ha, it would involve clearing approximately 0.5% of the forest compartment. On the other hand, some landings that will develop only a few truck loads can be smaller. On these smaller landings, a dedicated loader may not be efficient and the skidder will

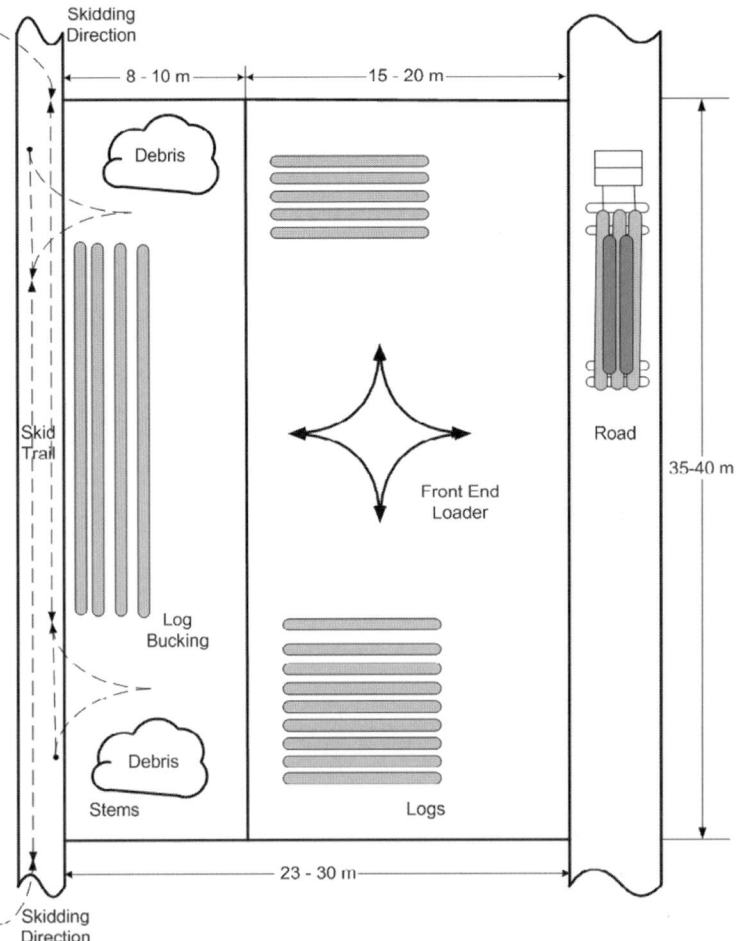

Fig. 6.15. Landing example where tree lengths are bucked on the landing, sorted, and loaded on trucks at road side by front-end tracked or wheeled loader

push and pile the logs as necessary to arrange log decks that can be loaded out later. A grapple-equipped skidder provides additional flexibility for moving logs around the landing.

Following completion of harvesting, landings should be restored so that proper drainage occurs to reduce soil erosion and runoff. Corduroy should be removed. Bark and landing debris should be distributed evenly across the landing to assist in stabilization. The site should be cleared of nonbiodegradable material, including oil and fuel drums and wire rope.

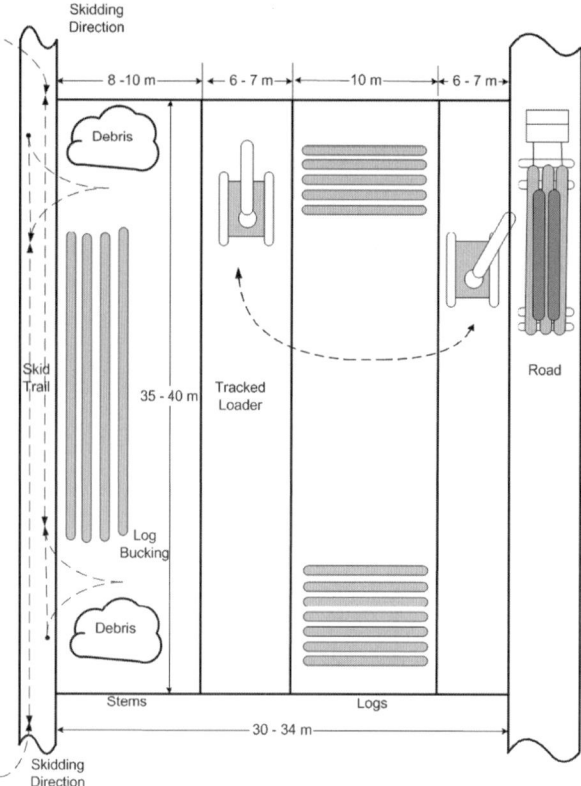

Fig. 6.16. Landing design for skidding tree lengths to landing, bucking on landing, with sorting by swingboom loader and roadside loading by swingboom loader

6.5
Spur Road Spacing

For skidders, forwarders, or agricultural tractors the most important factors which can be controlled are load size and distance. In general, the heavier the load that can be pulled or carried without excessive wheel or track slip, the more economical is the skidding operation. Similarly, the shorter the skidding distance, the more economical the skidding operation. However, the logging manager is responsible for both skidding costs and road construction costs. Placing the roads too close together, although good for production, is not good for overall costs. No discussion of log skidding is complete without an introduction to the spacing of roads to minimize the sum of skidding costs and spur road (feeder road) costs. Spur roads form the fingers of the road network

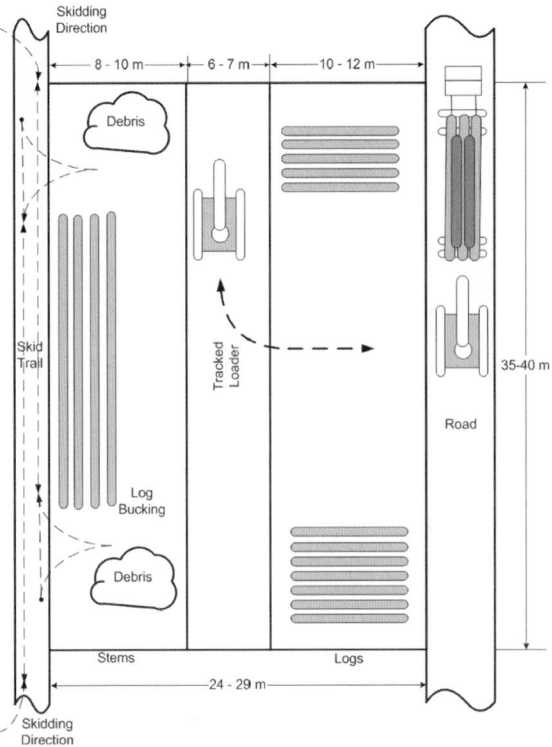

Fig. 6.17. Landing design for skidding tree lengths to landing, bucking on landing, with sorting by swingboom loader and on road loading by swingboom loader

penetrating to the landings to which skidding and forwarding systems deliver the harvested wood. When using a specific skidding or forwarding machine, there is a spur road density (meters per hectare) or a spacing which will result in the lowest combined cost of constructing the spur road and skidding or forwarding. This is called the optimum spur road density (ORD) or optimum spur road spacing (ORS). For skidding costs which are linear with skidding distance, it is attained when the travel portion of the skidding or forwarding cost equals the cost of building the spur road and maintaining it during the hauling period. Under ideal forest conditions on flat or gently rolling terrain where spur roads are straight and parallel, skidding or forwarding is carried on perpendicularly to the road and equidistantly on both sides, and the loads are offloaded where the road is reached, the average skidding or forwarding distance is one quarter of the spur road spacing. However, this situation rarely, if ever, occurs in practice. Sometimes a spur road may follow the border of a swamp, lake, river, or other topographic feature, so skidding or forwarding is done from one side only.

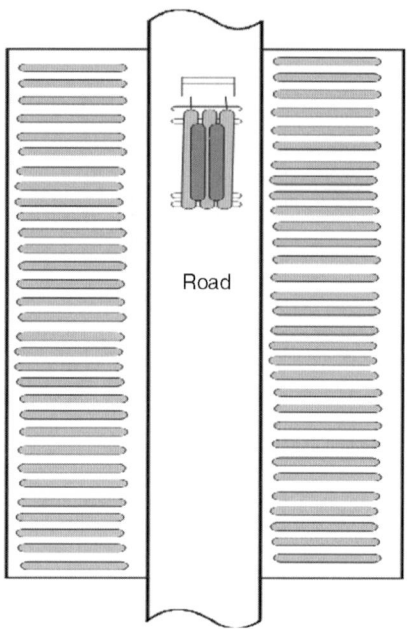

Fig. 6.18. Landing for loading shortwood stacked by forwarders from thinnings in planted forests

---

6.5.1
Optimum Spur Road Spacing

While spur road density, expressed in meters per hectare, is easier to use in calculating spur road cost per cubic meter, spur road spacing is the more practical guide for the logging engineer laying out a spur road network in a forest.

The ORS may be found with the formula

$$\text{ORS} = k \sqrt{\frac{40RL}{qct(1 + p)}},$$

where ORS is expressed in meters, R is the cost per kilometer of constructing and maintaining the spur road, L is the average skidder or forwarder load in cubic meters, q is the quantity of wood harvested, expressed in cubic meters per hectare, c is the operating cost per minute of the skidder or forwarder, including the operator, t is the time in minutes for the skidder or forwarder to travel 1 m loaded and return 1 m empty, k is a correction factor, with a normal value between 1.0 under the ideal conditions when skidding or forwarding is done equidistantly on both sides of the spur road and 0.71 (or $\sqrt{0.50}$) when skidding or forwarding is done from one side only – it is also used in situations where the spur roads are winding, meet in junctions, or terminate as dead-end roads,

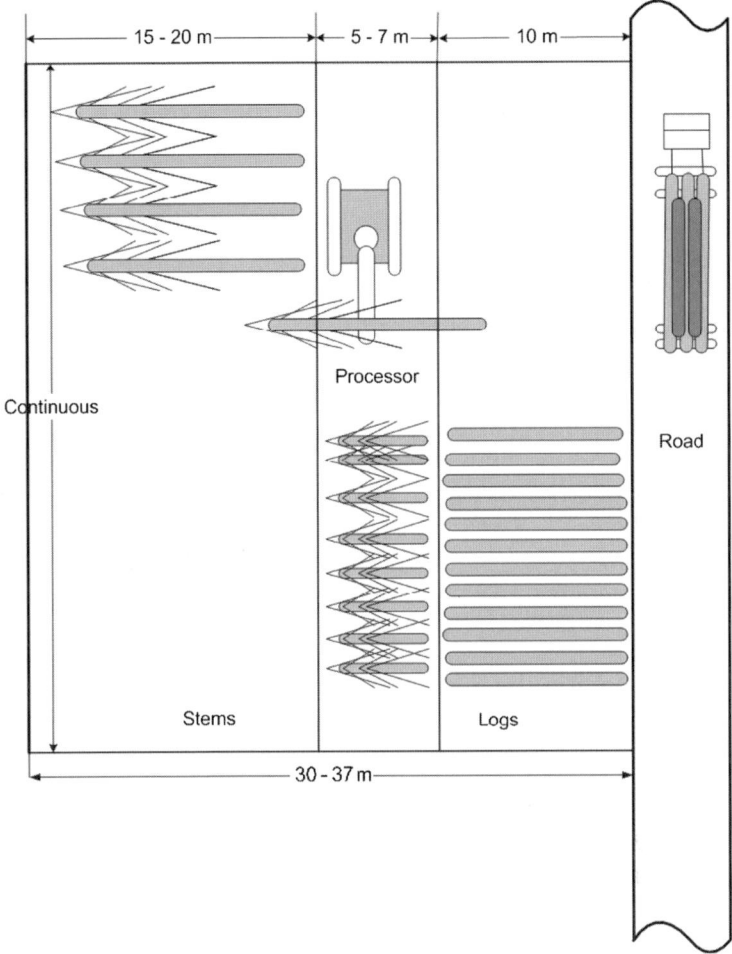

Fig. 6.19. Continuous landing design for tree length skidding to landing, processing on landing, and loading truck loading by rubber-tired swing loader on thee truck road. Frequently used in clear felling in planted forests

and p is a correction factor, normally with a value between 0 and 0.5, to be used in situations where skidding or forwarding trails, i.e., the strip roads, are winding or do not end at the closest point on the spur road, or where an allowance is made for delays along the route due to low-bearing soils, hang-ups, and so on.

The spacing distance derived with the formula may be considered only as an approximate value because of the imprecise values of several of the factors in the formula. For example, if the formula gives an optimum spacing of 400 m, a spacing between 350 and 450 m will give quite satisfactory results. This allows some leeway in locating spur roads to avoid obstacles that might increase the cost of constructing the road. An examination of the ORS formula will show that quadrupling the quantity of wood harvested per hectare will halve the spur road spacing; this will (1) require twice as much road to be built but at half the cost per cubic meter and (2) halve the skidding distance and therefore the traveling portion of the skidding cost, thus bringing about an overall reduction in the logging cost.

Example 1: What is the optimal road spacing if the road construction cost is $10,000 per kilometer, the average skidder load is 4 m³, the volume to be harvested is 30 m³/ha, the operating cost is $0.80 per minute, and the skidder travels 5 km/h unloaded and 3 km/h loaded? Skidding is equidistant from each side and a correction factor of 0.2 is used to account for delays.

The time to travel 1 m empty and to travel 1 m loaded is 60/5,000 + 60/3,000 = 0.032 min per round-trip meter.

The ORS is then

$$\text{ORS} = 1.0 \sqrt{\frac{40 \times 10000 \times 4}{30 \times 0.8 \times 0.032 \times (1 + 0.2)}} = 1318 \, \text{m}.$$

Example 2: What is the optimal road spacing if the road construction cost is $10,000 per kilometer, the average forwarder load is 14 m³, the volume to be harvested is 130 m³/ha, the operating cost is $1.20 per minute, and the skidder travels 3 km/hour unloaded and 3 km/hour loaded? Forwarding is equidistant from each side and a correction factor of 0.2 is used to account for delays.

The time to travel 1 m empty and to travel 1 m loaded is 60/3,000 + 60/3,000 = 0.04 min per round-trip meter.

The ORS is then

$$\text{ORS} = 1.0 \sqrt{\frac{40 \times 10000 \times 14}{130 \times 1.2 \times 0.04 \times (1 + 0.2)}} = 865 \, \text{m}.$$

The total cost per cubic meter in example 2 (Fig. 6.20) is a minimum at a road spacing of 865 m, but the total cost per cubic meter changes little between 500 and 1,400 m, even though the skidding cost and road costs are quite sensitive to change. This has important environmental and tactical implications. If it is environmentally desirable to minimize road construction, the roads can be spaced toward the upper end of the range (i.e., 1,400 m) without having much effect on total cost. But increasing road spacing has other environmental implications. If fewer roads will be built, the skid trails will be longer and there will be more trips per skid trail, increasing the potential for rutting and compaction. On the other hand, if there is a shortage of skidding equipment, or a risk of a shortage in the operating season for the skidders, then shortening the

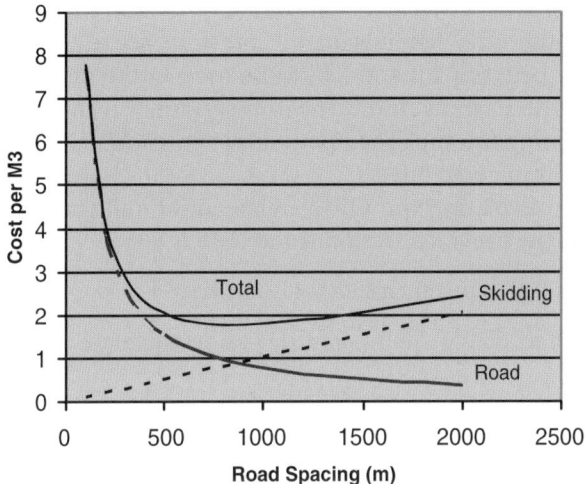

Fig. 6.20. Road cost, skidding cost, and total cost per cubic meter as a function of road spacing for the data in example 2

skidding distance toward the lower end of the range (i.e., 500 m) will greatly increase skidding productivity while not increasing total production cost. And, at the shorter end of the road spacing range, there will be fewer trips per skid trail as the skid trails will be shorter.

6.5.2
Optimum Spur Road Density

Having determined the ORS as in the previous section, the ORD may be found with the formula

$$ORD = \frac{10000}{ORS},$$

where ORD is in meters per hectare and ORS is in meters.

The ORD may also be found directly with the formula

$$ORD = 50\sqrt{\frac{qct \times 1000TV}{RL}},$$

where ORD is in meters per hectare, q is the quantity of wood harvested in cubic meters per hectare, c is the operating cost per minute of the skidder or forwarder, including the operator, t is the time in minutes for the skidder or forwarder to travel 1 m loaded and return empty, T is a correction factor, normally with a value between 1.0 and 1.5, to be used in the same situations as the factor k in the ORS formula in the previous section, V is a correction factor, normally between 1.0 and 2.0, to be used in the same situation as the factor p in the ORS formula in the previous section, R is the cost of constructing and maintaining the spur road per kilometer, and L is the average skidder or forwarder load in cubic meters.

Example for calculating the ORD: What is the optimal road density for example 2 from the previous section if the ORS is 865 m?

$$ORD = 10000/ORS = 10000/865 = 11.6 \text{ m/ha}.$$

If the ORS is not available, then the ORD can be calculated as

$$ORD = 50 \sqrt{\frac{130 \times 1.2 \times 0.04 \times 1000 \times 1.2}{10000 \times 14}} = 11.6 \text{ m/ha}.$$

6.5.3
Average Skidding Distance

Under the ideal forest situation the average skidding or forwarding distance is found from

$$ASD = \frac{2.5 \times 1000 TV}{ORD} \text{ or } ASD = TV \times \frac{ORS}{4},$$

where ASD is the average skidding or forwarding distance in meters, ORD is in meters per hectare, ORS is in meters, and T and V are correction factors as defined earlier.

Example for calculating the average skidding distance: Given $T = 1.2$, $V = 1.0$, ORS = 865 m, and ORD = 11.6 m/ha, calculate the average skidding distance.

$$ASD = (2.5 \times 1.2 \times 1,000)/11.6 = 259 \text{ m, or}$$
$$ASD = (1.2 \times 865)/4 = 259 \text{ m}.$$

6.5.4
Spur Road Cost

The cost of constructing and maintaining a spur road during harvesting may be found with the formula

$$RC = \frac{R \times RD}{1000q},$$

where RC is the spur road cost per cubic meter, R is the spur road cost per kilometer; RD is the road density in meters per hectare, and q is the quantity of wood served, expressed in cubic meters per hectare.

Example for calculating road costs: Given RD = 11.6 m/ha, R = $10,000 per kilometer, and q = 130 m^3/ha,

$$RC = \frac{10000 \times 11.6}{1000 \times 130} = \$0.89 \text{ per cubic meter.}$$

6.5.5
Variable Skidding Cost

The cost of the travel portion of the skidding or forwarding operation may be found with the formula

$$TC = \frac{ASD \times ct}{L},$$

where TC is the travel cost per cubic meter of skidding or forwarding, ASD is the average skidding or forwarding distance in meters, c is the operating cost per minute, including the operator, of the skidder or forwarder, t is the average time in minutes for the skidder or forwarder to travel 1 m loaded and return 1 m empty, and L is the average skidder or forwarder load in cubic meters.

Example for calculating skidding costs: Given ASD = 259 m, c = $1.2 per minute, t = 0.04 m/min, and L = 14 m^3,

$$TC = \frac{259 \times 1.2 \times 0.04}{14} = \$0.89 \text{ per cubic meter.}$$

If spur road density or spacing is optimum, the cost calculated in this manner should equal the cost of the spur road and the lowest combined cost of the two operations will have been realized. If this condition is deviated from, the combined cost will be greater.

6.6
Environmental Considerations

Environmental considerations primarily include soil disturbance leading to erosion, soil compaction, and impacts on streams. A number of measures can be taken to reduce environmental impacts of skidding. Education, training, and supervision of the work force are important elements in reducing disturbance.

6.6.1
Surface Disturbance

Surface disturbance that may lead to erosion can be reduced by:

- Reducing the number of trips made over the same trail. If only a few trips are made over a skid trail, the slash will still be intact. If more than a few trips are made, it may be just as well to use the skid trail to its fullest before creating more trails. Using fewer trails concentrates the soil disturbance (and soil compaction) on a smaller percentage of the area rather than distributing it over a larger area.
- Controlling the season of use to the extent practical where distinct wet and dry seasons exist.
- Leaving slash on the skid trails. The slash must be in close contact with the soil or water flowing in wheel ruts will continue to erode even under a slash cover.
- Keeping gradients on trails as low as possible, particularly when vehicles are traveling loaded.
- Locating skid trails on higher, well-drained, or easier-to-drain locations. Avoid seeps or springs where soil strength is always at a minimum.
- Using skyline yarding on steeper slopes so that soil compaction and trail excavation is minimized.
- Avoiding conditions requiring the use of chains.
- Matching load size to trail and equipment conditions. Wheel or track slip increases markedly when load size is excessive.
- Releasing the load and crossing weak spots if they cannot be avoided before winching up the load and continuing to the landing.
- Reducing tire and track slip by increasing the traction capability of the skidder to pull a given log load.

For rubber-tired skidders traction can be increased by choosing larger tires and/or using wheel tracks. Traction capability for rubber-tired skidders is also increased by increasing the number of powered wheels.

Traction capability for tracked skidders is increased by maintaining as long a length of track in contact with the ground as possible. Flexible tracks and sprung suspensions permitting the road wheels to move independently are generally more effective in doing this. A longer, narrower track generally requires less slip to pull the same load than a shorter, wider track of the same track area.

- Reducing tire and track slip to pull a given log load can also be achieved by reducing the motion resistance of the tire or track by reducing tire or track sinkage. This can be accomplished by reducing inflation pressure or increasing track area.
- Avoiding gouging of the soil with the leading end of a skidded log by using an integral arch or sulky to lift the front end of the log. If an integral arch is used, choking the large end of the log increases skidder traction and lowers log skidding resistance. This also reduces the required track or tire slip to pull the load. However, choking the large end of the log also increases the rear-axle load or track loading.
- Placing adequate drainage in skid trails after harvest to avoid erosion. The interval or spacing depends upon gradient, soil type, and expected rainfall intensity and amount.
- Performing skid trail maintenance after harvest if there are deep ruts.

6.6.2
Soil Compaction

Soil compaction may reduce plant growth by reducing air voids and water permeability. It also promotes soil erosion by increasing surface runoff. Soil compaction is affected by soil type, moisture content, ground contact pressure, wheel load, vibration, and the number of passes. In general, the higher the ground contact pressure, the higher the degree of compaction at any depth. The higher the wheel load, the deeper the compaction effect will go. Most of the compaction occurs within the first ten trips. Compaction is usually least when the soil is driest, but significant compaction may still occur over a wide range of moisture levels.

Soil compaction can be limited by several methods:

- By minimizing soil compaction in the individual skid trail. This can be done by (1) reducing inflation pressures for rubber tires and increasing track area for tracked skidders, (2) reducing wheel loading through smaller loads or adjusting the load distribution, using lighter equipment, or increasing the number of axles, and (3) reducing the number of trips over the same trail.
- By concentrating the soil compaction on a smaller percentage of the total area using preplanned skid trails. This restricts equipment to fewer trails rather than distributing the compaction over a larger area. This also reduces the area for rehabilitation.
- By rehabilitating the site after harvest through soil tillage.

6.6.3
Activities Near Streams

- Avoid locating skid trails or landings in riparian zones. The riparian zone can be an effective filter to prevent sediment from erosion from reaching streams if it is left in an undisturbed state.
- Drain water from any trail, landing, or truck road into locations where water can be absorbed by the soil before reaching a stream channel. Revegetate all disturbed soil surfaces as quickly as possible. Slash or other organic matter can provide almost immediate protection if in contact with the ground.
- Where the probability is high that some excessive erosion may occur and be damaging to a stream, trap the sediment behind logs placed in contact with the ground and parallel to the slope contours. This will provide some sediment storage capacity until the eroding surface is protected by plant growth.
- Where intermittent streams must be crossed in the dry season, protect the stream banks by temporarily placing logs in the stream channel.
- Where streams with active flow must be crossed, use portable bridges. Portable bridges can be made from a variety of materials. Laminated wood structures are often used. Portable bridges for skid trails are installed by a grapple skidder or by winching the panels into place with a skidder or crawler tractor. When using a grapple skidder, the grapple skidder picks up the panel, backs the panel into place across the stream, and then lowers the panel directly on the stream bank.

6.6.4
Training, Incentives, and Supervision

Technical solutions may be ineffective without operator awareness, skill, incentives, and supervision. Even skilled operators may be unaware of resource problems and solutions, so on-site supervision and communication with skidding operators often are vital in controlling soil and water impacts. In many cases, a small adjustment by an operator can yield significant benefits. For example, timber cutters may not realize that when trees are cut and felled in random directions, maneuvering of skidders may cause unnecessary soil disturbance as well as skidding delays. Similarly, tractor operators should be directed to use winches or grapples to lift logs partially off the ground during skidding, and also to avoid using the tractor blade as a braking or maneuvering device.

Careful supervision is especially important when it is unavoidable to skid during wet weather or on weak soils. Close attention should be given to changing soil conditions, and skidding should be suspended if rutting or other disturbance causes unacceptable impacts.

Prior to closing the cutting block, supervisors should confirm that spur road maintenance has been performed.

Cable yarding is the transportation of logs from the stump area to roadside with one of the various cable yarding systems. Where ground skidding systems can feasibly operate, cable yarding systems are generally more expensive than ground skidding and require larger volumes of wood to be removed per hectare to be economical. On swampy, steep, or very broken ground, cable systems may be more efficient, particularly if the costs of roads and harvesting are considered together. The application of cable yarding systems may also result in lower surface disturbance and very low compaction. Cable yarding systems are divided into two classifications: highlead and skyline. Helicopter yarding is the transportation of logs from the stump area to a point at the roadside using a cable suspended beneath the helicopter. Helicopter yarding is a flexible system for yarding uphill or downhill and may compete favorably with cable systems with respect to delivered log cost in areas of high road cost and low log volume removals per hectare.

7.1
Highlead

In highlead yarding, the log is dragged from the stump to the landing (Fig. 7.1). If the log becomes stuck, the tightening of the hauling-in cable (mainline) exerts an upward lift to the log which helps free it. Sometimes the haulback line also is used to free the log and to provide temporary lift to move past an obstacle. This system is suitable for clear felling only. The yarding direction can be either uphill or downhill. Substantial ground disturbance can result from the dragging of logs with this system. The yarding distance is often less than 300 m. Highlead yarding is not permitted in many natural forests because of the need to selectively harvest trees while protecting the residual stand, and controlling ground disturbance and log breakage. While the yarding equipment may look and sometimes be similar, the highlead system is *not* to be confused with the skyline system (Sect. 7.2).

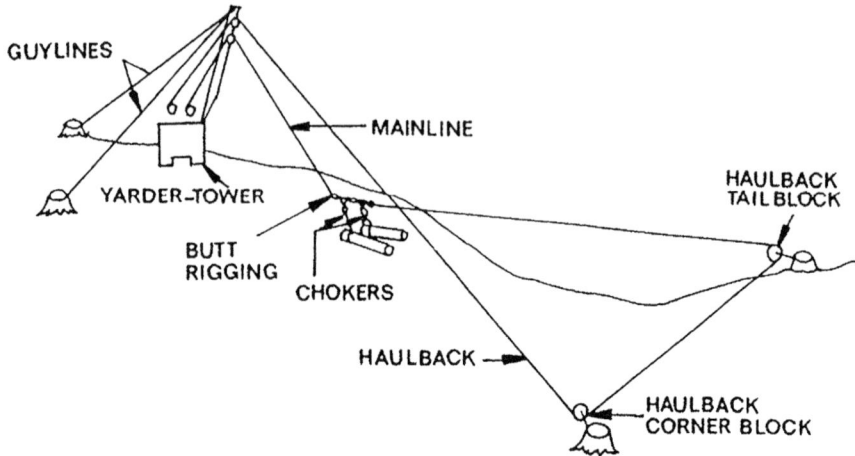

Fig. 7.1. Highlead

7.2
Skylines

In skyline yarding, one or both ends of the logs are suspended during transport. This improves yarding efficiency and decreases ground disturbance. The logs are fastened to a carriage and the carriage is pulled to the landing suspended from a wire rope cable known as the skyline. There are several types of skyline systems. The most common are the live skyline, the standing skyline, and the running skyline. Skyline systems of suitable power, with sufficient cable, and under proper topographic conditions can yard logs as far as 2,000 m. Skyline yarding is primarily used in plantation forests where large volumes per hectare are available for extraction.

With the live skyline, the carriage is pulled from the landing to the stump area by a line known as the haulback line. If the yarding is uphill and the slope is greater than about 20%, the haulback line is often not used and the carriage returns to the stump area under the force of gravity (Fig. 7.2). When the carriage reaches the stump area, the carriage is stopped by the applying a brake to the mainline and the skyline is lowered to the ground for log hook-up. After the logs have been attached to the carriage, the skyline is raised and the carriage is brought to the landing.

The standing skyline looks like the live skyline, except that with the standing skyline the skyline is not lowered during the yarding cycle. The standing skyline must be equipped with a carriage which can lower a line to the men in the woods (choker setters) so that the logs can be hooked (choked). When used

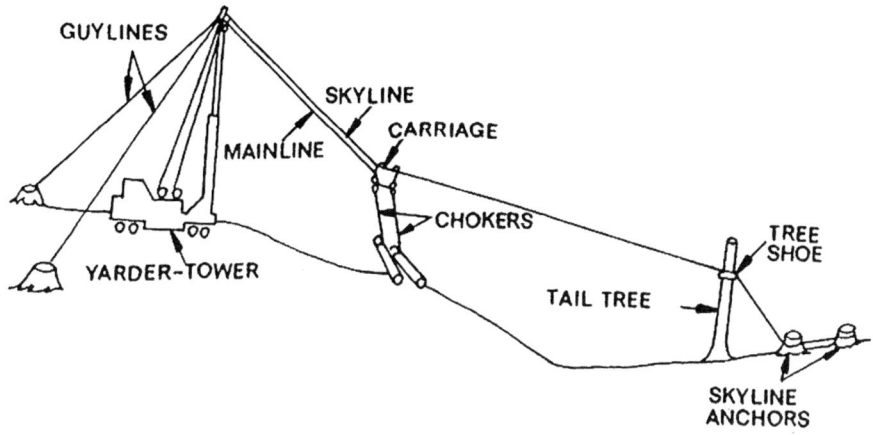

Fig. 7.2. Live skyline, gravity return

with a suitable carriage, either the live skyline or the standing skyline can be used to yard uphill, downhill, partial cut, or clear-fell. The standing skyline is probably the most common skyline. The running skyline is a special kind of live skyline (Fig. 7.3). With the running skyline, the skyline and the haulback line are the same line. When used with a suitable carriage, the running skyline can be used to yard uphill, downhill, partial cut, or clear-fell. For clear felling, grapples can be used. Most running skylines have a special mechanical or hydraulic linkage (interlock) between the haulback line and the mainline to recirculate power to make the system mechanically efficient.

7.3
Yarders

A yarder is a machine with one or more powered winches. A yarder can have between one and five winches (drums). A five-drum yarder would have a skyline, mainline, a haulback line, a slackpulling line, and a strawline (rigging) drum. Often a steel tower is attached to the yarder. The yarder can be mounted on a trailer or a self-propelled carrier such as a truck, tractor, skidder, tracked undercarriage, or rubber-tired undercarriage (Figs. 7.4–7.9). Sometimes the yarder is mounted on a sledge and a tree is used as a wooden tower (spar). Yarders can be classified by the maximum mainline pull: small (less than 100-kN pull), medium (less than 300-kN pull), and large (over 300-kN pull).

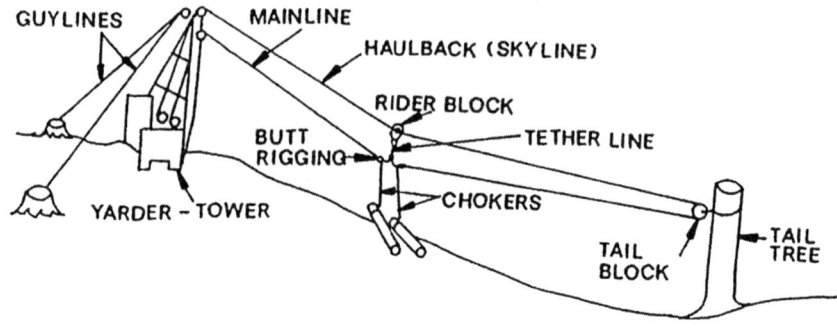

Fig. 7.3. Running skyline

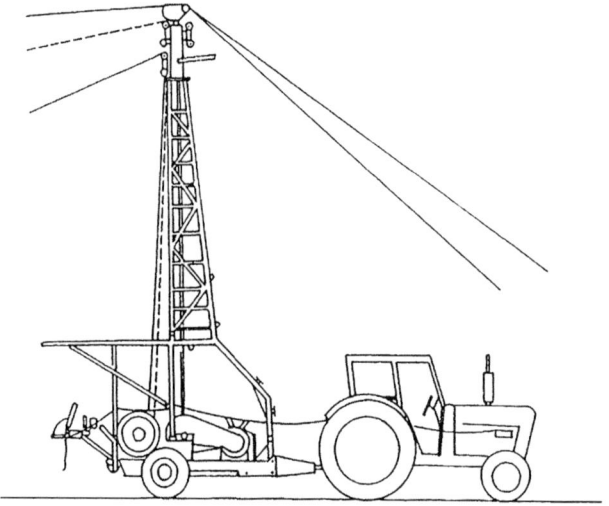

Fig. 7.4. Small skyline yarder, trailer-mounted

7.4
Carriages

The type of carriage used on a cable system largely determines its application. For the highlead system, a carriage is not used, and the logs are attached to chokers suspended from a collection of shackles and swivels called the butt rigging (Fig. 7.10). The butt rigging is only used for clear felling. There is no way to way to protect the remaining trees from being damaged.

Skyline carriages can be divided into two types: non-slackpulling (Fig. 7.11) and slackpulling (Figs. 7.12–7.16).

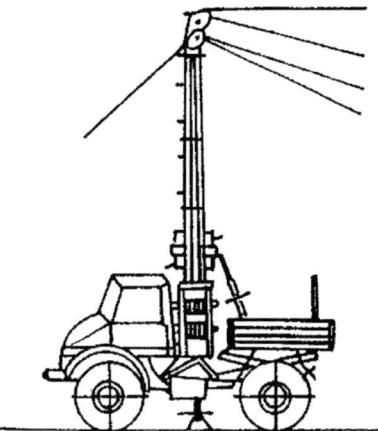

Fig. 7.5. Small skyline yarder, truck-mounted

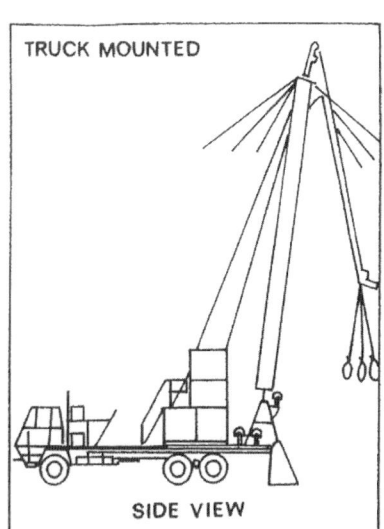

Fig. 7.6. Medium skyline yarder, truck-mounted

To be able to partially cut a stand while protecting the remaining trees, the carriage must allow a line to be pulled from it to where the logs lay. The pulling of the logs to the carriage is called lateral yarding. Carriages which have this ability are called slackpulling carriages. The most common slackpulling carriages on live and standing skyline either permit the mainline to be pulled through the carriage or drop a separate line to the ground. This separate line, sometimes called the skidding, tag, tong, or drop line, is stored on a drum in the carriage. The skidding line can be pulled manually, or the yarder can assist by using either the slackpulling line or the haulback line to pull a skidding line

Fig. 7.7. Medium skyline yarder, skidder-mounted

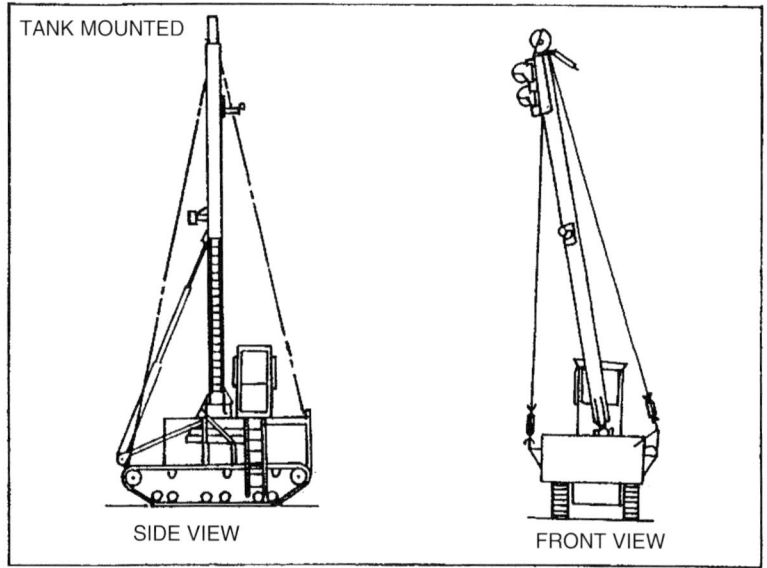

Fig. 7.8. Medium skyline yarder, tank-mounted

from the carriage. For some large cable systems, the carriage can have a self-contained motor which can power a slackpulling drum or can pull the mainline through the carriage to the ground. If either the yarder or the carriage assists in pulling out the cable for log hooking, the system is said to have a mechanical slackpuller. For downhill yarding or yarding at long distances, a mechanical slackpuller is necessary because the line is too heavy to pull

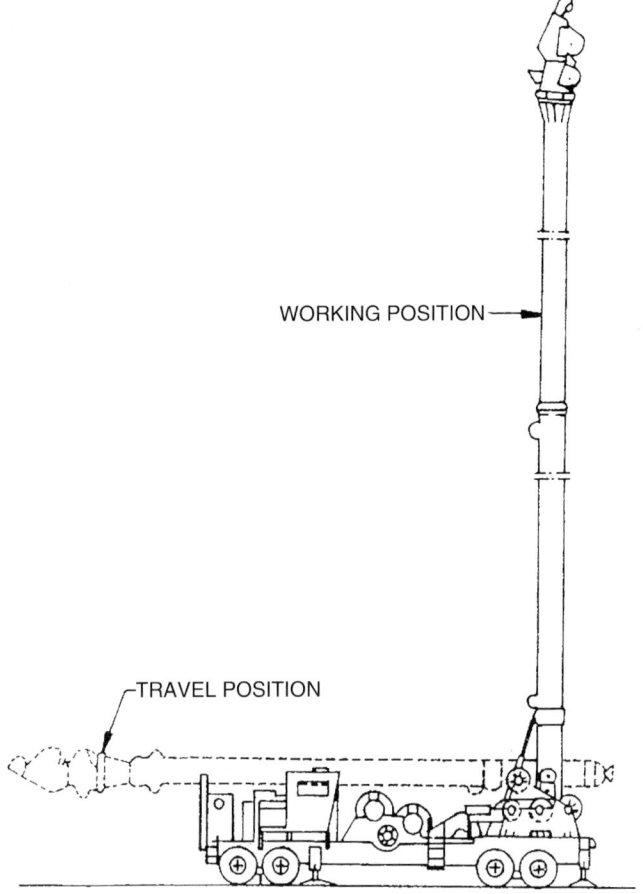

Fig. 7.9. Large skyline yarder, rubber-tired, self-propelled

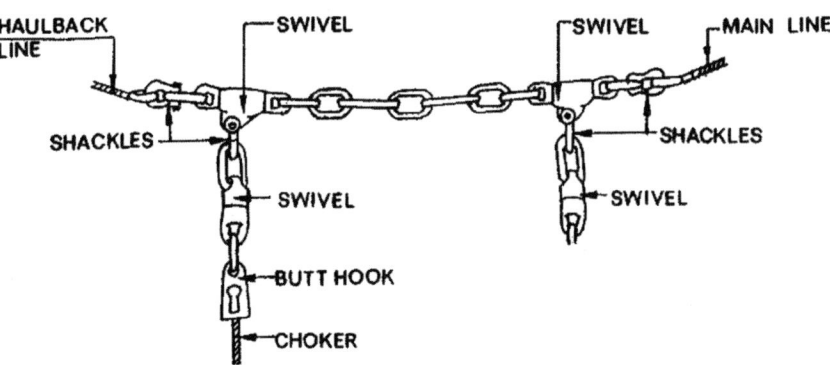

Fig. 7.10. Highlead butt rigging

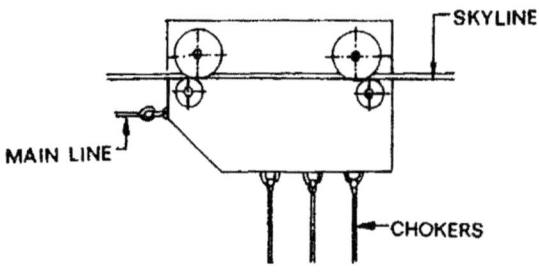

Fig. 7.11. Non-slackpulling sky-line carriage

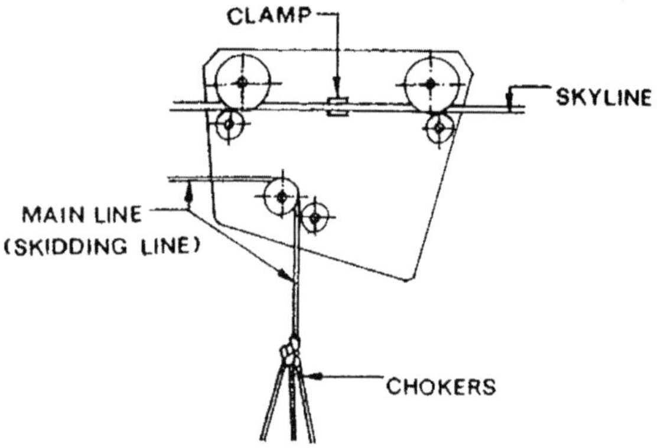

Fig. 7.12. Slackpulling carriage with skyline clamp

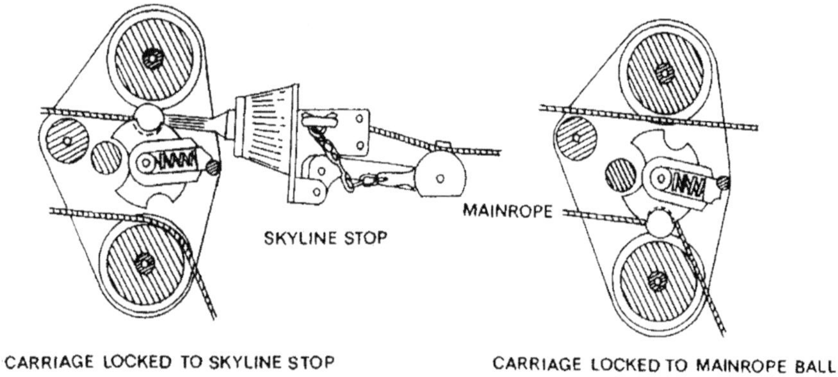

CARRIAGE LOCKED TO SKYLINE STOP   CARRIAGE LOCKED TO MAINROPE BALL

Fig. 7.13. Slackpulling carriage with skyline stop

manually. For a standing skyline, a mechanical slackpuller may be necessary to enable the choker setter to reach the line carrying the chokers.

To be able to pull in the logs to the carriage, the carriage must be held in location to prevent it from running up the skyline. This can be done with a haulback line or by clamping the carriage to the skyline. Skyline clamps can be engaged (set) in a number of ways. The clamp might be set by manually pump-

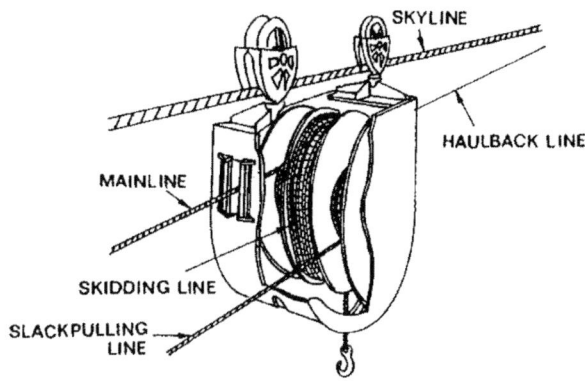

Fig. 7.14. Slackpulling carriage with haulback

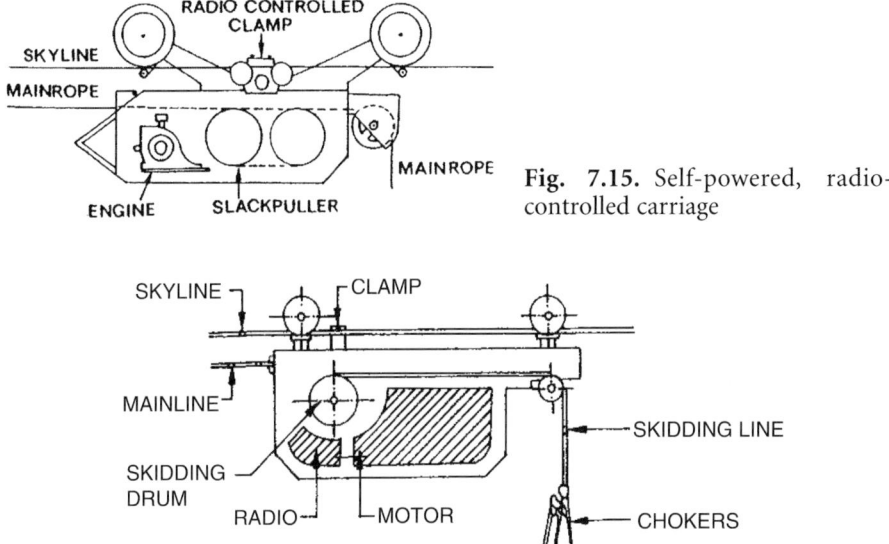

Fig. 7.15. Self-powered, radio-controlled carriage

Fig. 7.16. Self-powered, radio-controlled carriage with independent skidding drum

ing up a hydraulic cylinder, by physically engaging a movable stop on the sky-line, by using a remote radio-controlled clamp, or by activating a hydraulic clamp through a change in carriage direction. Similarly, clamps can be released automatically when the hook reaches the carriage or upon a radio signal. In unusual cases, a very simple carriage could be choked to a tree during lateral yarding and then released manually to yard the logs to the landing. For live and standing skylines which use a haulback line to maintain carriage position during lateral yarding, it is more efficient to have the log load locked into the carriage when the carriage is being pulled up the skyline to reduce fuel consumption and brake wear on the haulback line. Slackpulling carriages on running skylines (Fig. 7.17) can also be used to operate grapples instead of chokers to pull in logs (Fig. 7.18). Carriages which can pass an intermediate support provide the flexibility to yard over convex profiles while maintaining log suspension and reduced ground disturbance (Fig. 7.19).

7.5
Load Capacity

The load that a cable system can carry is limited by the lowest of four factors: the pulling capability of the yarder, the safe working tension of the skyline and mainline, the design of the carriage, and available tail trees or anchors. For

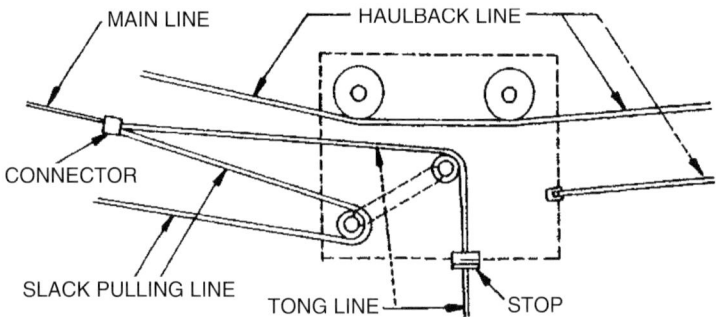

Fig. 7.17. Slackpulling carriage for running skyline

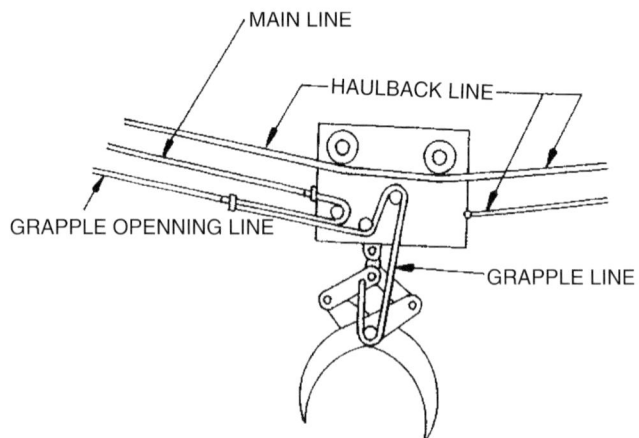

Fig. 7.18. Grapple carriage for running skyline

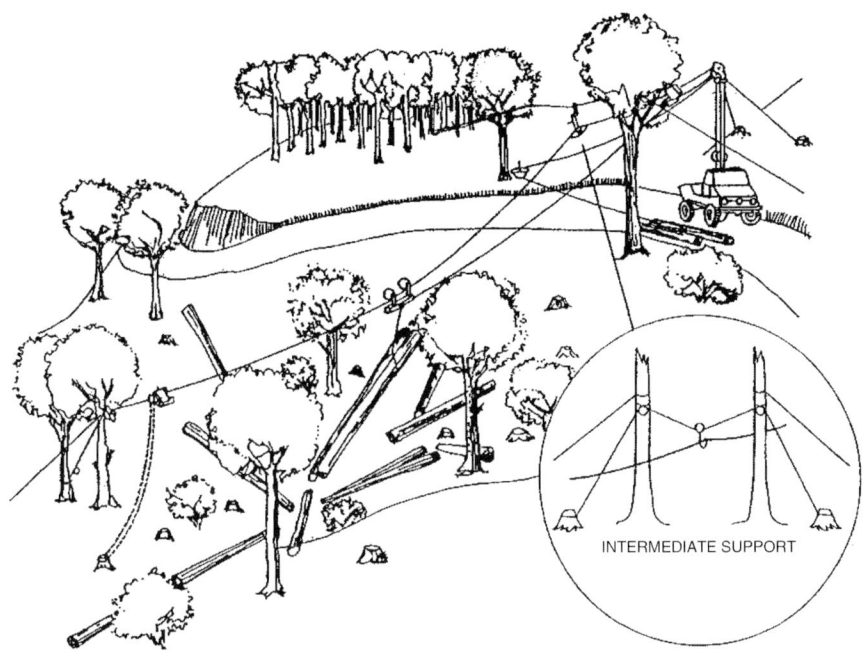

Fig. 7.19. Standing skyline with intermediate support

small skyline yarders the pulling capability by the yarder is often the most limiting factor. For larger yarders, the safe working tension of the skyline is often limiting. The breaking strength of new wire rope is divided by a safety factor to arrive at the maximum allowable tension. A factor of safety between 3.0 and 5.0 is often used to determine the maximum safe working tension of the skyline for logging planning purposes.

To determine the largest load that a skyline can support, one needs to know the profile of the terrain over which the cable will be suspended. The more sag (deflection) that is available, the greater the log load that can be supported by the skyline. Figure 7.20 illustrates this important idea. An analog method, known as the chain and board method has been widely used to estimate the available sag. Several computer programs have been developed to evaluate profiles for skyline logging. One widely used program is LoggerPC (Oregon State University).

In general, the larger ratio of the allowable skyline sag to the skyline length, the larger the allowable log load on the skyline. If logs are allowed to drag one end on the ground, a larger load can be carried although tension in the mainline increases owing to the log drag. One guide is that a skyline can support a dragging log load about 60% larger than it can fully suspend. Permissible log

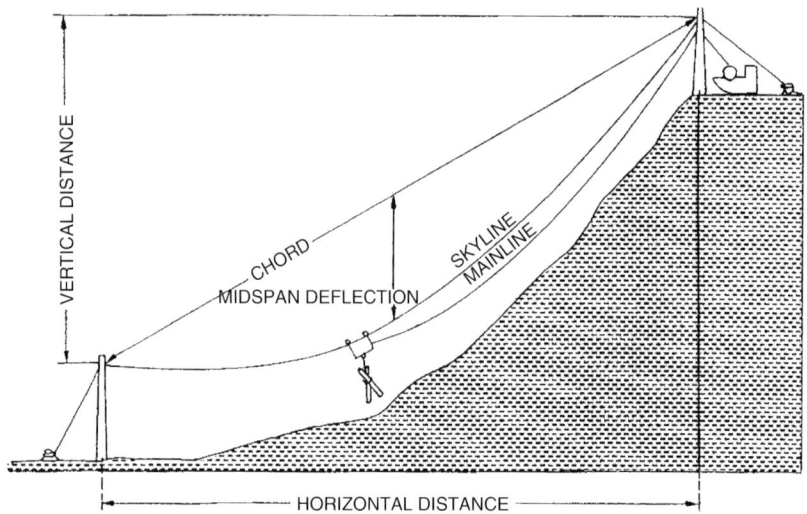

Fig. 7.20. Chord slope and midspan deflection of a skyline

loads for full and partial suspension can be determined from formulas, tables, and nomographs. Allowable log loads in tables and nomographs are often based upon the available sag (deflection) at middle of the span (midspan). Usually less than 5% midspan deflection is marginal, and more than 15% deflection is usually more than adequate to keep the skyline tension below its maximum allowable tension. Load-carrying capacity can be increased by using a larger cable, a higher strength cable, shortening the span, and increasing the height of the supports. The use of intermediate supports also can be used to increase load-carrying capacity.

Intermediate supports essentially reduce the effective span by dividing the overall span into shorter segments. The largest load that can be brought to the landing from the log pickup area will be limited to the smallest of the allowable log loads for the spans between the point of log pickup and the landing. Intermediate supports can be of several types: two-tree supports (Fig. 7.19), single-tree supports, and leaning trees.

7.6
Planning for Cable Yarding

In planning for cable operations the most important factors are (1) to select the appropriate equipment and (2) to verify that an adequate log load can be carried. To select the appropriate equipment you need to know if you will yard uphill, downhill, will make a partial cut or a clear felling, and the size of the timber. This will dictate the type of yarder and carriage. If thinnings are planned, it is essential to be able to hold the carriage at a location and to have lateral yarding capability. If downhill yarding is planned, the haulback drum must have adequate braking capability. Downhill yarding is to be avoided in thinnings owing to frequent hang-ups behind standing trees.

To know whether an adequate log load can be carried will require verification. A yarding network must be planned. Potential landings are identified on contour maps and the deflection checked. Landings are then connected by truck roads using grade projection to stay within the truck operating limits. Landings on a natural site should be at least 1.5 times the longest log length plus the road width. For a landing in an excavated site, the landing should be at least the maximum log length plus the road width. The locations of landings, guyline anchors, tail trees, skyline anchors, and intermediate supports should be verified in the field. Marginal profiles should be field checked for deflection. If adequate stumps are not available for anchors then either the landing or the skyline corridor must be moved or some type of artificial anchor must be provided. Artificial anchors include buried logs (Fig. 7.21), heavy logging equipment such as an old crawler tractor, and a variety of metal anchors, including tipping plates and pickets.

Yarding patterns for clear fellings are generally radial (Fig. 7.22). For thinnings, yarded by small, mobile yarders, parallel settings are common.

Sometimes, ground skidding equipment is used to bunch logs at a bench or the bottom of a slope and the skyline is used to yard (swing) the logs to the landing. In a few instances, the ground skidding equipment has been transported by skyline to the work area.

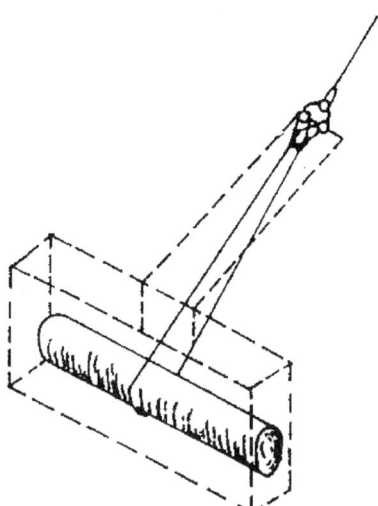

Fig. 7.21. Buried log for substitute anchor

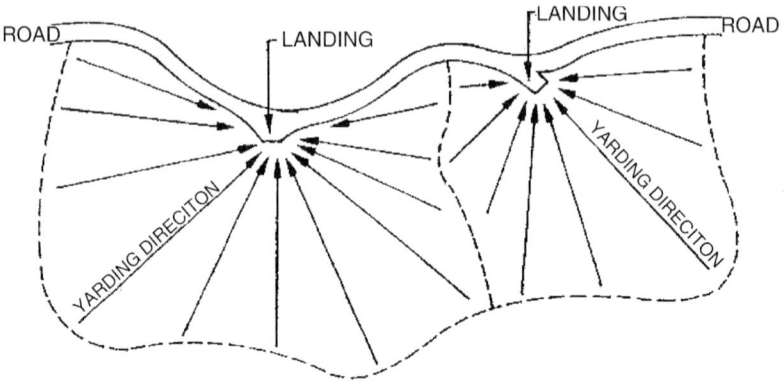

Fig. 7.22. Radial landings for skyline

7.7
Landings for Cable Yarding

Landings for cable yarding for clear-felling operations include the required space for the yarder and the loader. Trucks will need an additional spot to turn around (Fig. 7.23). The yarder will often remain in one spot with only its guyline shifted as necessary to best support the yarding direction.

The loader serves several roles as it clears the chute, segregates the logs into their sorts, and stores them safely on the landing, as well as loads the trucks. The loader operator must take care to load the oldest logs not just take logs from the top of the sorting piles, which may be the ones with the newest falling date. This can prevent losses due to staining and other decay. Often it is the loader operator who will track the log volume by the various logs sorts and is responsible for issuing the log tickets to the truck driver to show a transfer of ownership from the forest owner to the log hauler. The loader operator is also often responsible for keeping the landing free from the logging debris.

With the numerous activities that occur on the landing, yarding, sorting, delimbing, bucking, and loading, the landing can be a very dangerous environment to work in and adequate space is needed to ensure that the ground workers remain visible for the loader and yarder operators. The chaser is usually the only person who is on the ground on a cable landing and performs several activities on the landing. These include unhooking turns unless electronic chokers that are self-releasing are employed. The use of electronic chokers can eliminate one of the most dangerous aspects of the chaser's job as these devices can eliminate climbing on unstable log stack to manually release the chokers. Although the majority of the delimbing and bucking is

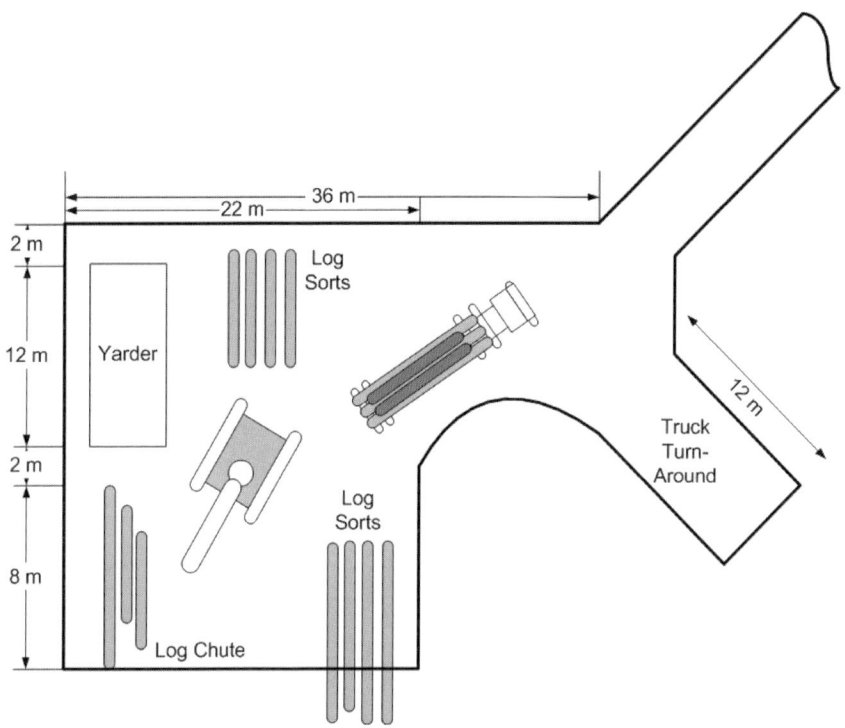

Fig. 7.23. An example of a landing for cable yarding

accomplished at the stump prior to yarding, some logs will need additional delimbing and bucking to meet mill specification and this is normally done by the chaser.

The size of the cable landing is dependent on the size of equipment. If a 30-m tower is needed, the landing will be larger than if a 12-m tower is used. The length of the chute where the logs are landed should be long enough to hold two-thirds of the length of the logs being yarded to the landing. If tree-length material is being yarded, with a length of 40 m, the chute may approach 25-m long. If a mechanized processor is used to delimb and buck, additional landing space will need to be provided. Another aspect that defines landing size is the number of sorts being produced from the stand. A larger number of sorts will require additional areas to allow the accumulation of full loads. The size of the landings can exceed 0.2 ha in situations where eight or more sorts are done.

The landing can see significant loader and truck traffic, and maintenance of drainage during the wet season is paramount for economic, safe, and environmentally sound forest operations. Mud accumulation will slow the pace of the chaser and lower the quality of his work as well as increasing the hazards from poor footing. The landing should be built using an out-sloped design to remove the water from the landing and ditches be maintained where the landing and road join. Where aggregate is not available, a landing may require the use of corduroy construction to prevent rutting of the subgrade on that part of the landing subject to traffic. Trucks should remain on the road to minimize the accumulation of mud and debris on tires that will contaminate the road surface accelerating its degradation.

Cable landings for thinnings involve smaller equipment removing lower volumes per hectare and can often utilize wide spots in the road. It is helpful to identify the candidate landings prior to road construction as the additional landing space can be built while the road is being constructed. Logs are often piled along the road for self-loading trucks or hauled to a storage site by three-wheeled loaders or small tractors with grapples to wait for self-loading or small grapple loaders.

7.8
Cable Yarding Operations

Landings and skyline corridors should be marked in advance of felling to guide cutters. Felling patterns should be made to facilitate yarding (Fig. 7.24). In natural forest, yarding through large crowns is difficult unless large equipment is being used. When uphill yarding, an alternative is to begin felling at the lower part of the unit and progress upward, yarding the logs as the cutting progresses so as to avoid yarding through tree crowns.

To keep landing and skyline road change time to a minimum, several sets of rigging are used. This permits tail trees, intermediate supports, and guyline stumps to be rigged in advance.

7.9
Cable Yarding Production

Productivity varies from 15 to 310 m^3 per 8-h shift, depending on the cable yarding system and machinery, yarding distance, and circumstances. Yarding productivity is also affected by the technical characteristics of the yarder, such as pulling power, load capacity of the mainline or skyline, cable configuration and deflection, and line speed. The work environmental factors also affect productivity. Such factors may be the size of trees, size, shape, and location of the stand, volume of timber per hectare, yarding distance, topography and length of slopes, terrain conditions, and weather conditions.

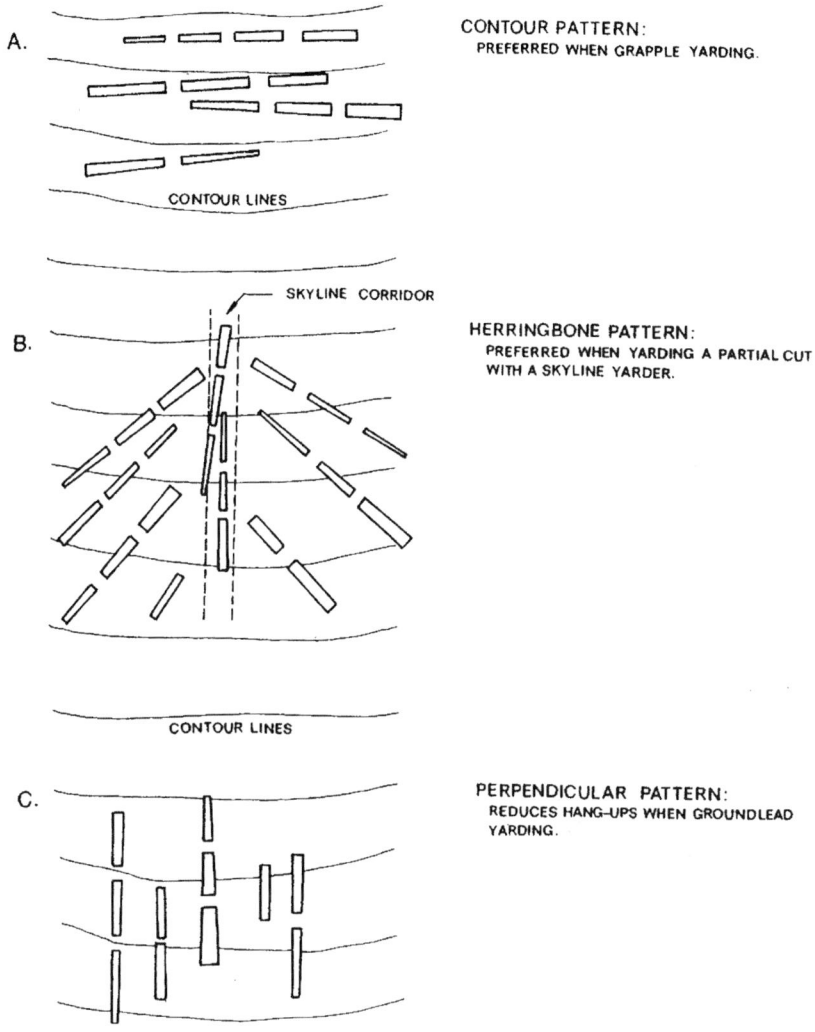

CONTOUR PATTERN:
PREFERRED WHEN GRAPPLE YARDING.

CONTOUR LINES

SKYLINE CORRIDOR

HERRINGBONE PATTERN:
PREFERRED WHEN YARDING A PARTIAL CUT
WITH A SKYLINE YARDER.

CONTOUR LINES

PERPENDICULAR PATTERN:
REDUCES HANG-UPS WHEN GROUNDLEAD
YARDING.

Fig. 7.24. Felling patterns to facilitate yarding

Productivity is affected by still other conditions, such as:

- Road network (location, standards, spacing)
- Terrain transport distance
- Lift and deflection
- Skill and organization of personnel
- Harvesting method (full trees, tree lengths, logs, etc.)
- Type of cutting (clear cutting, thinning, etc.)

7.10
Helicopter Operations

Helicopters are occasionally used to move logs from the stump to the roadside. Logs are either attached using chokers to hook at the end of a line (tagline) or grabbed using a grapple suspended from the tagline. Helicopter capacities range from about 1 t to more than 15 t. Helicopter operations are much more expensive than other logging methods, but provide the flexibility to pick up scattered high-value trees, react quickly to forest health problems such as salvage after fire or wind storms, and provide more flexibility in road locations. Helicopters can yard uphill or downhill and can be used for clear felling, selective cuts, or thinnings. For a given helicopter model, the allowable log load will vary depending upon temperature and altitude. Owing to the high cost of the helicopter, the yarding operation must be well planned and executed. Normally several sets of choker setters or grapple spotters are used to provide a continuous supply of available logs and one or more loaders and possibly multiple log landings are used to maximize helicopter productivity. Helicopter costs per unit volume vary with distance, elevation change per unit distance, wind, and access to the log load. Hooking logs from clear fellings is the least costly and hooking logs from a mostly closed canopy is the most costly. Helicopters have been used in several Southeast Asian tropical forests for a number of years. Stormy tropical weather can be a safety problem for helicopter operations.

Modern loading of logs is usually done mechanically. There are two general types of roadside mechanical loaders: swingboom loaders with grapples, and front-end loaders fitted with a log fork or grapple. Both types are mobile, either on tracked carriers or on rubber-tired carriers. Most of the machines can be used to load shortwood, tree length, or full tree. Some have limitations, for example, front-end loaders have difficulty loading shortwood less than 2-m long. Some shortwood forwarders have the capability of off-loading directly to the truck or trailer and thus saving a rehandling cost. When this is practiced, spare trailers, either semi or full type, are usually parked at the roadside.

When logs are large and the operation is too small to afford the capital outlay for modern loaders, there are many other ways of using machines to load vehicles. Tractors and winches of all kinds may be used to load material of sawlog length by lifting or parbuckling with or without the benefit of an A-frame, a gin pole, or a mast and swinging boom. Sometimes the winch may be mounted on and powered by the hauling vehicle through its transmission. In some cases, bulldozers are used to push large logs up skid poles onto the hauling vehicle.

In situations where logs are light enough to be handled manually, loading may be done by one or more men by lifting or rolling. This is particularly the case where wage rates are low and labor is plentiful.

8.1
Loading from Ramps

Loading from ramps is one of the oldest methods of loading large logs. Logs are loaded by rolling, pushing, or pulling from above or below the vehicle by hand, animal, or tractor power (Fig. 8.1).

Parbuckling with ropes or cables can be done by animals, tractors, or from a power takeoff from the hauling vehicle (Fig. 8.2).

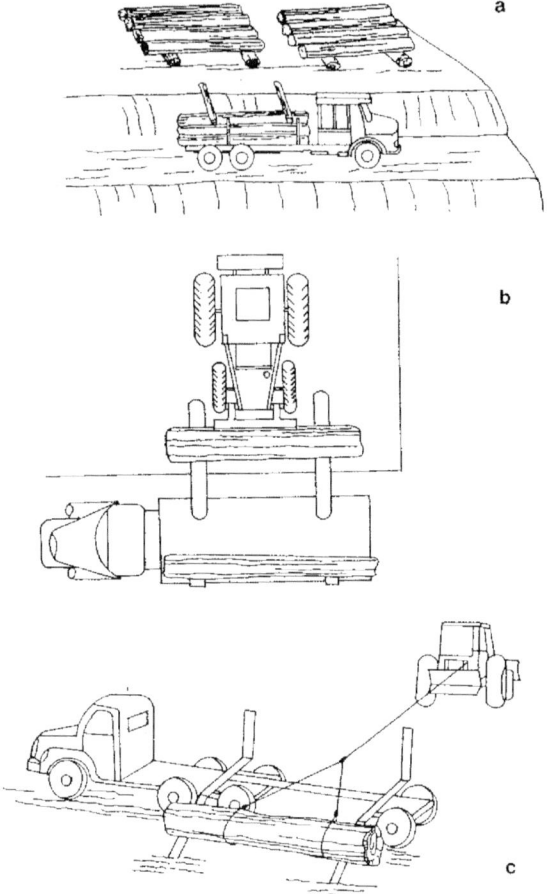

Fig. 8.1. Ramp loading. **a** By gravity for manual loading. **b** By gravity using a tractor. **c** By pulling by tractor

8.2
Loading with Swingboom Hydraulic Loaders

Swingboom hydraulic loaders are mounted on turn tables on their own mobile rubber-tired or tracked carriers (Fig. 8.3) or on the platform of a 6×4 or 6×2 truck of suitable capacity. Loaders on wheeled carriers are usually equipped with outriggers. When the loader is truck-mounted, the engine driving the hydraulic loader pumps is installed on the turntable. They may be fitted with a pulpwood grapple for loading shortwood or with a heel-boom grapple for loading tree lengths. Small knuckle-boom loaders may be mounted on the truck frame behind the cab of a truck or at some point farther back on a combination rig to form a self-loading, hauling unit.

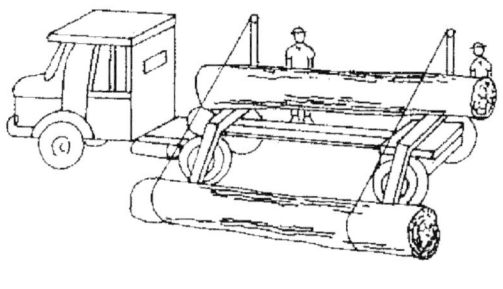

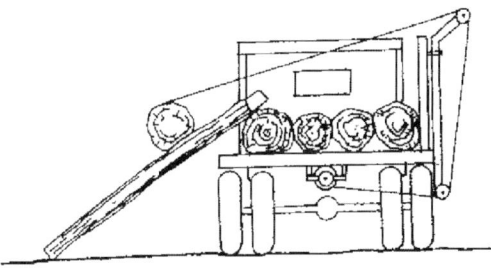

Fig. 8.2. Parbuckling using a truck-mounted winch

Table 8.1 gives some typical grapple payloads which various sizes of swing-boom hydraulic loaders can handle at various boom reaches.

8.2.1
Loading Long Logs or Tree Length

When loading long logs with high taper or tree lengths, the logs or tree lengths are piled perpendicularly to and with butts toward the road. The truck and the loader maneuver until the truck is facing the hauling direction and the loader, whether truck-mounted or carrier-mounted, is backed into position in front of and close to it. This takes about 3 min per load. The loader reaches out, grapples a load of tree lengths 3–4 m from the butt end, heels it against the heel boom, raises the entire load, swings it, and lays it down on the semitrailer by reaching back over the truck cab. If the loader cannot reach enough wood to complete the load, both the loader and the truck move to a new position. After the loading operation has been completed, the loader must be driven from the road at some point to allow the truck to leave.

When loading long logs without excessive taper which can be grappled from either end without log breakage, the loader can be positioned behind the truck.

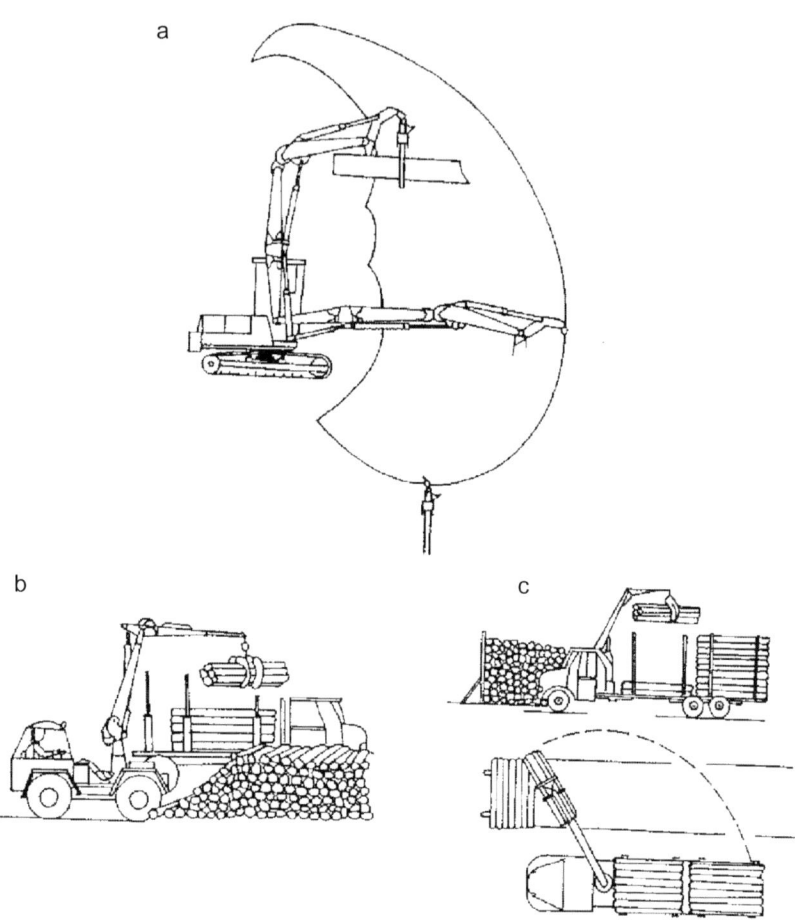

Fig. 8.3. Carriers for swingboom loaders. **a** Tracks. **b** Rubber-tired self-propelled. **c** Truck-mounted self-loading

Tests indicate that:

- The total fixed time per trailer load ranges between 5 and 7 min.
- The average grapple load is around 1.05 m^3.
- The average grapple cycle time ranges between 0.55 and 0.60 min, depending on operator efficiency.
- The loading rate is around 1.75 m^3/min, excluding the fixed time between 5 and 7 min.

Table 8.1. Some typical net payload capacities of knuckle-boom loaders

Type of loader mount	Maximum boom reach (m)	Typical net payload capacity (kg) at various boom reaches[a]		
		3 m	6 m	Maximum
Truck	6	5,000	2,350	–
	7.5[b]	8,500	4,100	3,000
Platform	9[b]	17,700	8,600	5,000
	6.5	2,200	900	850
	6.5	1,600	730	700
Four-wheel carrier	9	18,000	8,700	5,000
Truck frame behind cab	6	3,500	1,200	

[a]Net payload capacity is the gross capacity less the weight of the grapple.
[b]Counterweight with heel-boom grapple.

Loading time in minutes per truck-trailer load and loading cost per cubic meter may be then determined with the following formulas:

$$LT = FT + \frac{0.55L}{GL}$$

and

$$LCM = \frac{LT[C + c(1 + f)]}{60L},$$

where LT is loading time in minutes per truck-trailer load, LCM is the loading cost in US dollars per cubic meters, L is the truck-trailer load in cubic meters, FT is the fixed time per truck-trailer load in minutes, GL is the average grapple load in cubic meters, C is the operating cost of the loader per productive machine hour (PMH), including the carrier cost but excluding the operator cost, c is the direct wages of the operator per PMH, and f is the cost of fringe benefits expressed as a percentage of direct wages.

These loading times and costs do not reflect the time the loader may spend keeping the landing clear for skidders or the yarder, or for log sorting. Example for calculating loading production and cost: Calculate the loader production and cost given a truck capacity of 30 m³, a grapple load of 1.0 m³, a loader cost of $50 per hour, a fixed time per truck-trailer load of 6 min, an operator cost of $4 per hour, and fringe benefits of 40%.

$LT = 6 + (0.55 \times 30)/1.0 = 23$ min;

$LCM = \{23 \times [50 + (4 \times 1.4)]\}/(60 \times 30) = \$0.71$ per cubic meter

8.2.2
Loading Shortwood

Knuckle-boom loaders for handling shortwood have a boom and a pulpwood grapple and can be mounted on trucks, trailers, or wheeled carriers. Logs, the lengths of which approximate the maximum allowable load width, such as 2.5 m, are usually loaded crosswise on the truck or trailer as this provides the most compact form of load as well as easy unloading by dumping or pushing off. For this type of loading operation, the wood is usually stacked one or more rows along the roadside. The loader sits at the side of the truck, grapples the wood, and transfers it to the truck or trailer. It is a straightforward operation – the reverse of a forwarder offloading at the roadside.

The loading rate will vary with the size of the loader and grapple and the proficiency of the operator. The loading operation has (1) a fixed time per load to position the truck, pull away when loading has been completed, and apply the self-tightening load binders, and (2) a variable loading time per cubic meter.

As soon as one truck has pulled away, another may take its place, thus losing a minimum of loading time.

A typical knuckle-boom loader and 1.2-m^3 grapple can transfer 2.5-m wood to a trailer at the rate of 1.2–1.5 m^3/min.

8.3
Front-End Loaders

Front-end log loaders are mounted on a crawler-tractor chassis with some suspension changes or on a four-wheel-drive articulated chassis. All are equipped with a log fork with or without extension arms 50–60-cm long to increase lifting height and a "kicker" to assist in removing logs from the fork at high lift. Crawler-mounted log loaders range from 15,000 to 20,000 kg and wheeled log loaders from 15,000 to 35,000 kg. The horizontal skid tines of the log fork range in length from 150 to 200 cm depending on the size and weight of the machine.

Front-end loaders are used normally to load material of saw-log length and longer, including tree lengths which are loaded lengthwise onto the hauling rig. They may also be used to load 2.5–3-m logs lengthwise, and 2.5-m logs crosswise by loading the truck or trailer from the rear, but the practice is not recommended.

Front-end loaders travel between the log or tree-length pile and hauling rig carrying the loaded fork in the air. This requires that the ground, for best loading performance, be level, firm, and cleared of stumps and other entangling debris. For this reason the roadside landing is often bulldozed – an additional expense not encountered when heel-boom loading with knuckle-boom loaders sitting on the road. When the ground is soft, loading of tree lengths must be done with heel-boom loaders fitted with a special grapple.

Tree-length loading rate with front-end loaders depends on several factors: size and horsepower of the machine, proficiency of the operator, lifting height, conditions of the landing, distance between pile and trailer, and piling direction in relation to the road. Some tests show that:

- There is little difference in loading rate between wheeled and crawler machines of the same power rating if adequate traction exists.
- The fixed time per load (waiting for the truck to be prepared to receive logs and to drive away after loading has been completed, rearranging load, etc.) averages around 5 min.
- The average grapple load in cubic meters is 1.2% of the loader gross kilowatt rating.
- The average loader cycle time, excluding fixed time, ranges between 1.50 and 1.75 min.
- The average loading rate, disregarding fixed time, is around 0.75% of the loader gross kilowatt rating, when expressed in cubic meters per minute.

Loading time in minutes per truck-trailer load and loading cost per cubic meter may be found with the following formulas:

$$LT = FT + \frac{1.6L}{GL}$$

and

$$LCM = \frac{LT[C + c(1 + f)]}{60L},$$

where LT is the loading time in minutes per truck-trailer load, FT is the fixed time in minutes per truck-trailer load, LCM is the loading cost per cubic meter, L is the truck-trailer load in cubic meters, GL is the average grapple load in cubic meters, C is the operating cost of the loader per PMH, c is the direct wage of the operator per PMH, and f is the cost of fringe benefits expressed as a percentage of direct wages.

8.4
Unloading

There are many methods of unloading trucks and trailers. Much depends on the circumstances. When unloading into open water, shortwood is usually end-dumped, side-dumped, or pushed off with a bulldozer-like "pusher" with long arms and a pusher plate. When unloading on a lake, the same procedure may be followed, or the wood may be unloaded and piled with the same type of equipment as used in the loading operation and at approximately the same cost per cubic meter. Tree lengths or full trees are not usually unloaded for water transportation unless with a stiff boom or other crane in bundle-sized packages. Saw logs, tree lengths, and full trees may be unloaded and stored with a front-end loader, a stiff-boom crane and grapple, or a bridge crane, or unloaded onto a slasher or sorter deck with a pusher-type machine or a powered pulling device for immediate further processing. They may also be pushed off onto the ground to be slashed and hauled away or dumped into open water from a landing. Large powerful front-end loader types of machines capable of picking up and carrying an entire trailer load of tree length or full tree wood are in wide use.

9.1
Trucks

The productivity of truck transport is affected by the load size and the number of round trips (turns) made per day. To be effective, the truck should have a short terminal time (time for loading and unloading) and a reasonably high traveling speed. In most cases, truck transport starts from a landing. However, when the terrain is gentle and the soil strong enough, truck transport can be started from the stump site.

Transport distances may vary from some kilometers up to hundreds of kilometers. A product of low value, such as fuel wood, can be transported economically only over relatively short distances. Valuable roundwood, however, can be transported over much longer distances. In some countries, a suitable transport distance for two round trips per day is about 80 km. With poor road conditions only one round trip per day may be possible over such a distance.

9.1.1
Equipment

The choice of a suitable truck depends mainly on the size and form of the wood transported, transport distance, load-bearing capacity of roads and bridges, as well as conditions of loading and unloading. The choice of truck is also affected by possibilities for proper service and the availability of spare parts.

For timber transport, the trucks can be equipped with a flat bed, or steel body, and timber bunks with load-supporting stakes (stanchions) and binders for the load. Sometimes they are also equipped with a loading device. When choosing suitable truck units (rigs) for different transport conditions, selection must be made between the following alternatives:

- Two, three, or more axles
- Single, dual, or bogie wheels
- One, two, three, or more driving axles
- Load space on the driving unit or separate one-axle wheel attachment, separate bogie, semitrailer, or full trailer

When choosing a truck with a suitable load space, the common lengths and weights of the wood species to be transported have to be known. After that, the bearing capacity of the roads and bridges affect the choice. Restrictions on total allowable weights and the minimum allowable distance between the outer axles as well as the minimum allowable axle weight and the total length of the truck unit vary from country to country. Some types of timber trucks and their typical load capacities are presented (Fig. 9.1), where the first number expresses the number of wheels (or dual wheels), and the second number is the number of driving wheels (powered wheels):

- A single two-axle truck, also called a straight truck or four-wheel truck (configuration 4×2 or 4×4; load capacity 8–12 m^3).
- A single three-axle truck with double or bogie wheels at the rear end (6×4 or 6×6; 10–15 m^3).
- A two-axle truck, also called a truck-tractor, with a separate wheel attachment which is fastened to the log load (truck usually 4×2, with wheel attachment 20–25 m^3).
- A three-axle truck with a double-axle semitrailer, which may be equipped with a flat bed for shortwood or with a steel body and timber bunks with stakes for longwood (usually 6×2 or 6×4, 30–35 m^3).
- A three-axle truck with a connected three-axle full trailer (usually 6×2 or 6×4; 35–40 m^3).

From the alternatives illustrated above, the single two-axle truck (4×2 or 4×4) as well as the single three-axle truck (6×4 or 6×6) are commonly used in many tropical countries. Often they are used as "terrain trucks" as they are capable of transporting 5–6-t loads in terrain or along poor cattle-cart roads with deep ruts, and they can be easily turned in narrow places.

A separate bogie with a short beam may be used with the three-axle truck (6×4 or 6×6) to form a truck unit. Before loading, the bogie is placed behind the truck according to the length of the load. Then it is fastened to the longest log to be loaded, and the log is fastened to the truck. After that, the loading is carried out. Before driving empty, the separate bogie has to be lifted onto the deck by the truck's winch or by the truck-mounted loader.

Fig.9.1. Various truck and trailer configurations. (Courtesy Kantola and Harstela 1988)

A special type of semitrailer for long-log transport has a long boom under the timber bunk (Fig. 9.2). The distance of the semitrailer behind the truck tractor is determined by the length of the log load. During log transport the telescopically adjustable boom is hooked to the rear end of the truck-tractor frame. The tractor and semitrailer act as an articulated unit with less off-tracking than a standard truck-tractor with a semitrailer. Before driving empty, the semitrailer is lifted by the truck-mounted loader or other machine onto the

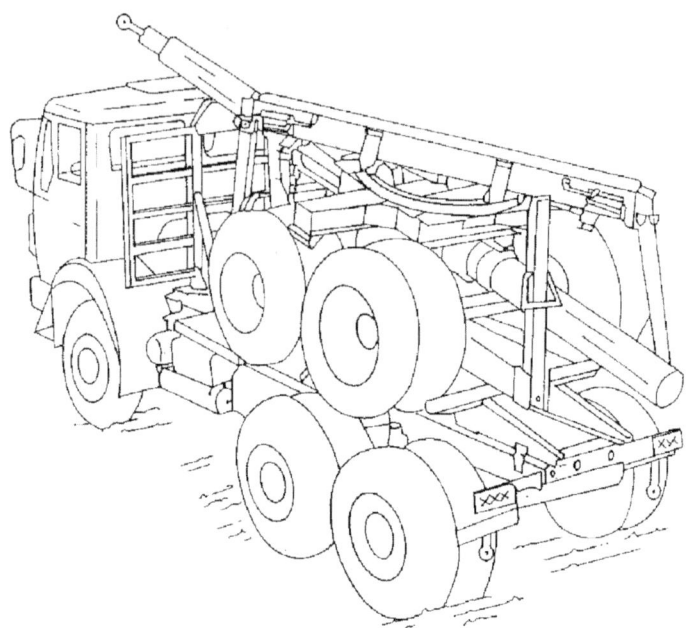

Fig. 9.2. Long-log truck with self-loader and trailer, piggyback. (Courtesy Kantola and Harstela 1988)

tractor. In this way the driving speed of the empty truck can be increased, traction is increased, tire wear is reduced, and the truck is easier to turn at the landing. With some designs, the semitrailer can be loaded onto the truck-tractor, without the use of a loader, by unpinning a hinge connecting the reach to the truck frame and backing the semitrailer against a ramp. Unloading the semitrailer at the landing is done by releasing the brakes on the trailer and letting it back off from the truck-tractor. A three-axle truck (6×2 or 6×4) with a full trailer is used on good roads and for long transport distances. The trailer may have two, or three, or even four axles. It may carry loads of 35–40 m³. Fully loaded it may weigh up to 50 t, depending upon local road regulations. A full trailer may also be used behind a semitrailer, or several full trailers may be used one behind another, when the roads and traffic regulations allow it. Other alternatives are also available. For instance, a four-axle timber truck with a two-axle semitrailer may be used on very good roads and for long transport distances. It may carry loads of up to 45–50 m³.

The log bunks are made of steel, and the load is supported by stakes. When using a semitrailer, the bunk of the truck-tractor is equipped with a turning plate (fifth wheel), to allow the load to turn according to the movements of the truck. If a flat bed is used, the bunks are mounted on that.

The stakes are installed at each end of the bunk. Integral stakes which have a safety tipping mechanism can be released while standing opposite the unloading side. The load is bound at least by two binder chains tightened around the log load.

9.1.2
Transport Costs

The yearly costs of truck transport consist of fuel and lubricants, maintenance, repair and service, tires, capital costs, and other fixed costs. In tropical countries, the share of capital costs and other fixed costs may form about half of the total annual costs. Fixed costs can be estimated using Table 2.3. To reduce these expenses, the truck should have enough productive hours per year. The daily costs consist of the capital costs of the truck, which also depend on the yearly amount of transport work, in addition to the daily operating costs. The daily costs per wood unit depend on the load size and the number of round trips per day. The size of the load is limited by the maximum load capacity and the condition of the truck, by the bearing capacity of the roads and bridges, and by other existing restrictions. It also depends on the climatic conditions and the vertical alignment of the road. The time needed for a round trip depends on the loading and unloading time, together with the driving speed and the distance of transport. The costs per load are derived by dividing the daily costs by the number of loads per day. Therefore, the size of the load is significant: larger loads are more economical than smaller ones.

Technical factors may limit the size of the load. A technical restriction may be a weak bridge on the road, or poor weather conditions, such as heavy showers or a rainy season. If a rainy season limits the use of trucks for many months per year, the fixed cost of a heavy truck may be too high. In such a case, smaller trucks are more economical.

Operating costs, including maintenance and repair, may be cheaper for a lighter truck, but calculated per cubic meter and kilometer they may be higher than for heavy trucks. The driving speed of heavy trucks on poor roads is low, consumption of fuel high, and need of repair extensive; therefore, heavy trucks may not be suitable on poor roads.

The road speed of a timber truck is mostly affected by the quality of the road, the number of adverse gradients, the radius of curves, suitability of the truck, and the competence of the driver. The costs of truck transport are affected by these speed factors. For example, when driving on a public road at a speed of 40 km/h, the costs per cubic meter per kilometer may be only half of those when driving on a forest road at a speed of 10–15 km/h. Consequently, the proper construction and maintenance of forest roads is important.

Legislation may set restrictions on the axle weight, width of the truck, length of the load, and thereby on the load size of a truck unit. When more axles are used in a truck unit, larger loads may be allowed. However, the total weight of a loaded vehicle on different kinds of roads and bridges may be restricted.

The transport of fresh wood is more expensive than that of dried wood; however, the transport of recently felled timber may be essential to avoid deterioration. The transport of longwood is usually less expensive than the transport of shortwood. An exception is the transport of longwood of poor form which limits the tonnage that can be loaded on a vehicle. Large variations in the need for repairs are caused by the care of the driver. The productivity of log transport is affected by driver skill. If the driver owns the truck, he is more likely to be careful. Excessive delays in transport result in high interest costs and disturbance in other operations. A skillful driver can influence the service life of his truck and at the same time diminish costs and the danger of accidents.

To organize the operations economically and to maintain the roads accordingly, truck transportation should be planned and coordinated. Poor roads should be used in good weather conditions and better roads at other times. When several trucks are used on the same narrow roads, traffic synchronization may be needed to avoid waiting times at loading sites or passing points. Scheduled speeds must be followed by the drivers. Registration and reporting devices and activities should be arranged.

9.1.3
Variable Tire Inflation

The use of radial tires and reduced tire inflation pressure is one method that can be used to increase the season of operation, improve traction, reduce vehicle operating costs, reduce road maintenance, reduce the depth of required road surfacing, and improve environmental performance of forest roads. At reduced tire pressure the footprint of the radial tire will increase (Fig. 9.3). The longer footprint reduces shear stresses for powered tires as well as for all tires during braking, improving traction and reducing the tendency to washboard. The longer footprint also increases the bearing area of the tire so that vertical stresses are reduced (Fig. 9.4).

The objective is to match tire inflation pressures with specific operating conditions defined by speed, terrain, load, and road surface strength. At lower pressures, impact loads to the road surface are reduced and washboarding and rutting are reduced. With less rutting, surface erosion from roads is reduced. Reports of crushed rock savings of 25% have been reported by reducing tire pressure from 690 to 345 kPa. Under severe road roughness conditions, travel speed can actually increase owing to reduced truck vibrations, and reduced tire pressures have been reported to smooth and flatten existing washboards.

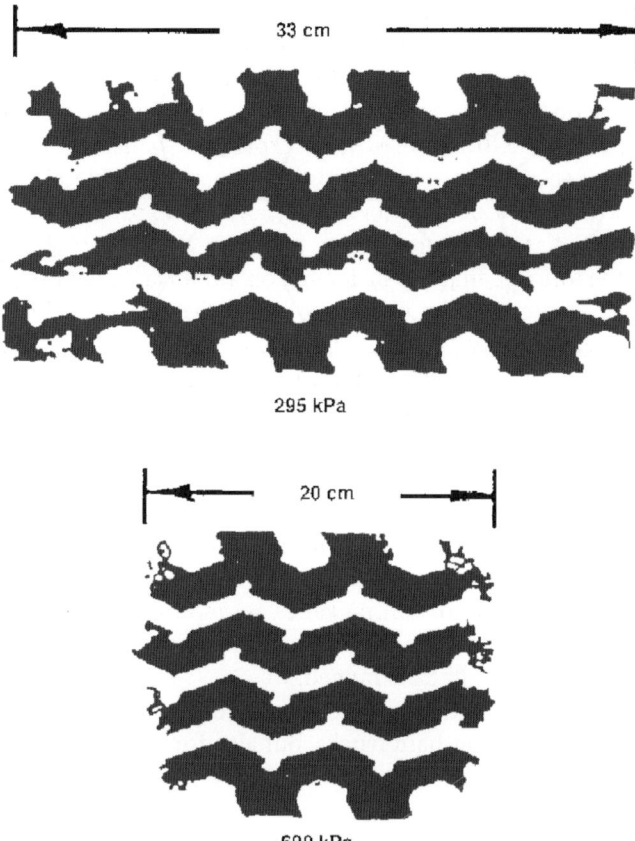

33 cm

295 kPa

20 cm

690 kPa

Fig. 9.3. The footprint of a radial tire will change size with different inflation pressures

Many radial-type truck tires can be safely operated at pressures 50% less than on-highway tire pressures if speeds are reduced to 60 km/h or less. Tire pressures can be adjusted manually when trucks enter the forest road system or trucks can be equipped with compressors, airlines, and controls that permit air pressure to be adjusted while vehicles are moving. On-board truck inflation systems can increase truck purchase costs by 10–20%. Costs for adjusting tire inflation, whether manually or automatically, must be balanced against savings in road surfacing costs, road maintenance costs, truck maintenance costs, and driver productivity. For situations where trucks only operate within the forest at speeds of 60 km/h or less, the tire pressures can be permanently reduced without purchase of additional inflation systems or controls. However, the benefits of reduced tire inflation require the use of radial tires as bias ply tires

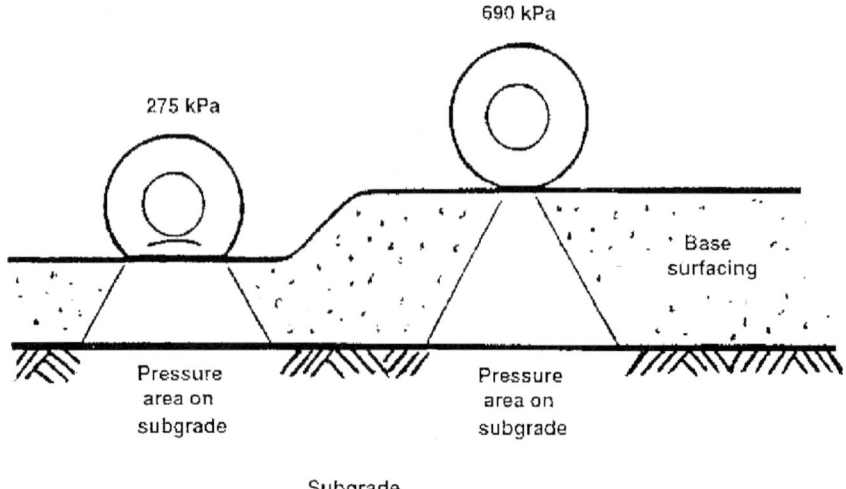

Fig. 9.4. Reduced pressure in a radial tire requires less surfacing to distribute loads to the subgrade

deflect differently under reduced pressure. Reduced tire pressure should be considered not only for log-haul trucks, but also for dump trucks.

9.1.4
Truck Maintenance and Repair

To transport wood safely, efficiently, and economically, the maintenance and repair of the trucks must be arranged. This must be planned and organized according to the scale of the operation and the degree of utilization of the trucks. A supply of spare parts should be obtainable, fuel stations and repair shops should be available, and capable machine mechanics should be trained.

The maintenance program of each truck should be given to the driver. His responsibility is to know when the truck has to be maintained at a fuel station and for which cases a repair shop is needed. He should be capable of carrying out certain maintenance operations and small repairs which do not require special tools.

Trucks require daily service. According to the check list a certain number of critical points must be checked. The level of the engine oil, gear-box oil, water level in the cooler, lights, and air pressure of the tires must be checked before starting out. Checking under the truck may reveal an oil leak or possible breaks in the spring joints. This is usually the best time for refueling too.

At the end of the day, damages and faults must be reported in order to get them repaired before the next shift. The repair service should be arranged so that small running repairs can be made immediately after the day shift of the driver.

Periodic maintenance is done according to the check lists on the basis of the kilometer reading. They also indicate which operations must be done in a repair shop or at a fuel station. The operator's manual specifies the tasks needed at different intervals. They may be lubrication, change of motor oil, motor filter, and hydraulic filter, or air/water removal from the container of the brake system. Cleaning or replacement of certain parts, or other tasks, may be needed after a certain number of kilometers.

Preventive maintenance in a repair shop or fueling station is needed at certain intervals. Critical parts of the truck may be replaced, and in this way the standing time of the truck considerably reduced. Urgent repair is needed when a breakdown has occurred or when the driver has located some sudden fault through an unusual noise during driving. If he is not able to make a temporary repair, he should stop the truck immediately and seek repair personnel.

9.1.5
Driving and Safety

A truck is an expensive forest machine. It must be driven along difficult forest roads with large loads, as well as on public roads in heavy traffic. If the truck is not handled correctly and maintained properly, truck transport may be difficult, expensive, and dangerous.

Drivers must be trained. Drivers should be provided with the knowledge about the function, structure, usage, maintenance, and repair of a truck unit. If road traffic regulations and legislation exist, they should be known. Drivers should also be familiar with the documents needed in road transport.

Through training, truck drivers should learn the skills of loading, driving, maintenance, and repair. They should also be able to follow the basic safety measures in wood transport. They should learn positive attitudes to orders given, and to accept responsibility in their work. During working hours the drivers should be carefully supervised. They need adequate food, rest, and sleep. No alcohol and no passengers should be allowed. On long distances, a half-way checkpoint for control of the vehicle, refueling, and meal breaks should be arranged. When loading, the driver is responsible for the correct distribution of the load. He must also check that the truck is not overloaded. After loading, he has to check that the stakes are well placed, and tools and accessories are secured, to prevent them from falling or bouncing. Before starting out, the driver must check the brakes of the truck and trailer. It is especially important that the braking power of the trailer is strong enough. If the brakes of the trailer are weak, the trailer may dangerously push the truck-tractor when driving downhill, or the brakes may fade when sudden braking is needed.

The quality of soil affects tire traction. On dry clay, tire traction ceases at a lower speed than on a sand-gravel surface. On wet soil, especially on wet clay, tire traction is readily lost. On soft surfaces the rolling resistance is high. When a vehicle is starting, it needs to overcome the resistance due to inertia, as well as rolling resistance and grade resistance. The inertial resistance may be equivalent to climbing as much as a 10% grade. Therefore, the landing should be placed on soil with good bearing capacity and if possible the loaded vehicle should depart downhill rather than uphill.

Table 9.1 gives an idea of the engine power required for a loaded tractor with a semitrailer weighing 40 t to maintain a constant speed on a gravel or laterite surface climbing different road gradients. Fuel consumption increases rapidly at higher speeds.

A correct driving speed must be applied. One rule is that trucks should descend a hill in the same gear as they would climb it under the same loaded condition. Trucks which must descend long downhill grades should be equipped with engine brakes to prevent overheating of service brakes. The driver must drive defensively and know his stopping distances. Braking too suddenly can cause jackknifing. The driving speed has to be lowered to match the road conditions and special care has to be applied where roads are poor. Table 9.2 gives an idea of the stopping distance as a function of road gradient and speed on a gravel road.

The driver should stop and check his vehicle at the top of long downhill grades and before entering a public road. For steep uphill grades he should be in the correct gear before he starts up the hill. Shifting gear on a steep grade should be avoided. Before entering a public road, the driver should check his load. If the load needs to be tightened, it must be done to secure a safe drive at higher speeds.

Table 9.1. Speed as a function of engine flywheel power and grade for a 40-t truck on a gravel surface

Grade (%)	Speed (km/h)					
	50 kW	100 kW	150 kW	200 kW	250 kW	300 kW
0	20	37	51	62	72	81
+5	6	12	18	23	29	34
+10	3	7	10	14	17	21
+15	2	5	7	10	12	15

Table 9.2. Example of minimum stopping distance on a gravel surface as a function of speed and grade for a truck with service brakes on all wheels

Grade (%)	Stopping distance (m)		
	20 km/h	40 km/h	60 km/h
0	22	54	97
−5	23	59	107
−10	25	65	121
−15	27	75	143

9.1.6
Extended Terrain Transport

9.1.6.1
Agricultural Tractors with Trailers

In some cases the same vehicle can be used both off the road and for longer-distance transport. This method is called extended terrain transport. In planted forests in flat terrain with a dry, even surface and good load-bearing capacity, extended terrain transport can be carried out with an agricultural tractor. The tractor can be started directly in the stump area and continued on the road. In the case of a 10–15-km road transport, a farm tractor with a semitrailer may be used. For longer distances a truck is required. In this way the terrain transport can be extended economically from the stump area to the public road without unloading and reloading at the landing. Sometimes, conditions on the public roads are so poor that a tractor with a semitrailer is a viable alternative for road transport. Agricultural tractor and trailers or semitrailers are generally used for log transport from plantation forests and short-distance transport in tropical forests. Often they are used for short-distance fuel wood transport.

A farm tractor with a strong engine and a suitable semitrailer may also be used in extended terrain transport. A semitrailer with a load capacity of 3–15 t places a part of its load on the rear axle of the tractor, increasing the traction for the tractor. If a semitrailer with driving wheels is available, it increases the transport capability of a farm tractor in the terrain even more. In the case of long-distance road transport, larger tractors are preferable. Full semitrailers cannot be used off the road.

Depending on the vehicle, and the terrain and road conditions, there may be large variations in the productivity of extended terrain transport. To give some idea of the influence of the transport distance, Table 9.3 gives productivity estimates for a large farm tractor with a semitrailer. The size of the load is assumed to be 6 m^3. The volume of annual transport production is estimated

Table 9.3. Estimated productivity of a large farm tractor pulling a semitrailer

Transport distance (km)	Trips per day	Productivity (m³/day)
1–5	3	18
6–15	2	12
16–25	1	6

by multiplying the daily volume of production by the annual number of effective transport days. The cost per cubic meter (without road and other additional costs) is calculated by dividing the daily costs of the tractor transport by the daily volume of wood transport. When the economy of a tractor and a truck are compared, it may be shown that the daily costs of a farm tractor are 25–30% less than those of a truck. However, the most suitable solution can be found when the load size and number of turns per day for the transport roads in question are known. The load size greatly affects the productivity and costs, and most for longer transport distances; therefore, vehicles with a large load capacity should be preferred.

The influence of terrain is often decisive. Good productivity can be obtained only on dry, flat terrain with an even surface and good load-bearing capacity. Even small gradients may lower productivity; a maximum manageable adverse gradient for a loaded vehicle may be 6–8%. When driving loaded downhill, on a dry surface, a gradient of 10–15%, or a lateral slope of 5–8%, may be allowed. A wet surface may cause a tractor or truck wheel to spin even in flat terrain.

9.1.6.2
Off-Road Trucks

Trucks used in extended terrain transport are called off-road trucks or terrain trucks. They should be capable of moving even on soft ground and be strong enough to carry heavy loads. Off-road trucks are normally all-wheel drive with large-diameter, wide-base, low-pressure tires. In Southeast Asia one popular off-road truck is a converted 6×6 military transport truck. In other cases, trucks designed for off-road construction or mining have been fitted with a fifth wheel, log stakes, and a pole trailer. These trucks can operate on very steep grades, poor running surfaces, and cross swampy ground. Off-road trucks are used between the log landing and a transfer yard along the main road. In some situations, they are used to haul directly from the landing to the mill or riverside. The economics of using off-road trucks should consider differences in road standards and the extra cost of reloading the logs if a transfer yard is required, and the costs increase in the hauling season. In areas of expensive road construction, where steep grade road locations could avoid large excavation, landslides, and potential erosion, the use of off-road trucks could be a viable alternative to limit environmental impacts. In some cases, landing size can be reduced as the off-road trucks can jackknife in the road to be loaded by a front-end loader. An additional advantage of off-road trucks is the ability to remove logs cut from the road right of way that normally would have to wait until road construction is complete. Often roads where off-road trucks are used cannot be used by on-road vehicles. Tire penetration can be as much as 60 cm. One consideration in the selection of a particular model of off-road truck is the availability of maintenance and parts. Off-road trucks from construction and mining equipment may share many of the same parts as the road construction equipment, so standardizing on a particular manufacturer has maintenance advantages in remote locations.

9.2
Water Transport

In the tropics transportation of round timber by water has always been important. In Latin America, the Amazon, Paraguay, and Parana rivers with their tributaries, are the most important. On the west coast of Africa, from Côte d'Ivoire to the People's Republic of the Congo and the Democratic Republic of the Congo, there are many usable streams and lakes. And, in the Far East, from Bangladesh to Indonesia and Vietnam, timber-producing tropical countries use water transport by driving, rafting, or on barges and ships, and seagoing rafts. Water transport is expected to remain important in such areas, although concerns about effects on water resources may result in some restrictions.

The different methods of transportation by water include extraction from mangrove and tidal forests, from swamps, and from seasonally inundated forests. Water transport is also used in combinations with truck transport for logs from tropical high forests, particularly from remote areas or for export purposes. Manually guided rafts are slow, going down the stream at a speed of 1 or 2 km/h and not traveling at night. Where tides exist, the rafts are held up by the incoming tide. The raft sizes vary within wide limits, from two to three logs of 4–5 m³ guided by one man in swift water, to 80 or 100 logs of about 300 m³. These rafts, descending with the stream, are too slow to be an economic success and continue only where tugboats cannot pass. This is usually because of the shallow water which may be deep enough to keep the timber floating but not deep enough for the tugboats, which are not specially designed for the tropics and have a much greater draft than the timber.

For safe and regular rafting of large volumes of timber, tugboats must be used. Traveling day and night, with rafts sizes up to 2,500 m³, and stopping only at the coast at the time of adverse tide water, tugboats, working in teams of two, can double or triple the speed of manually controlled rafts, and can be guided safely around river bends and between sandbars.

The buoyancy of floating timber depends upon moisture content (with most species having a density of more than 600 kg/m³ at 15% moisture content). Heavy green timber should not be rafted for long distances without being attached to some type of floater.

All timber to be rafted, floating or not, should be given time to season in order to increase buoyancy and, therefore, the duration of floating. The height above water of the emerging part of the log diameter is only an indication of the wood buoyancy at a given moment, and does not indicate the rate at which the water is being absorbed and how long its floatability will last when it is immersed. Some tropical woods pick up the water faster than others. Some keep their floatability for years and others lose it after a few weeks or months.

Rafts of light timber, riding high on the water, need less towing power and travel faster than those of heavy floating timber or mixed rafts of floaters and sinkers which lie low in the water. Nonfloating timber, called sinkers, can be kept up in a raft by the buoyancy of some lighter wood species, by bamboo bundles, by various palm species, or even by empty gasoline and oil drums.

When timber species with a lot of sinkers are logged, the available light woods may be insufficient to carry all the sinkers. Where the logging site is located within the range of reverse-tide waters, the floaters can be towed upriver and reused, if dried between the raftings. If the timber has to be rafted from locations beyond the reach of the tidal water, to which the floaters cannot be towed back, the floaters can be replaced by empty, sealed gasoline or oil drums brought in shiploads. The drums are placed end to end, in one or two rows, alongside the log and on both sides, and fastened to it with wire. Six to ten logs are bound together to make up the permitted width of the raft. Another system of rafting nonfloatable timber is by the suspension of the logs from outriggers laid across low barges. In forest regions with an excess of floating timber and few sinkers to be transported, the assembling and binding of mixed rafts can be avoided by rolling sinkers from the shore onto the rafts across the floaters and wiring them tightly together so they will not shift.

Sometimes rafts are pushed down the rivers instead of towing them. Logs fastened solidly into rigid rafts of floating timber, or of combined floaters and sinkers, are much easier to guide and to control by pushing than by towing, provided the length and the width of the rafts are properly adapted to the worst river bends and bottleneck passages. By mounting a pushing blade onto the bow, a tugboat can be used independently for both towing or pushing rafts.

The pilot has the raft right under his eyes and all boat maneuvers can be more easily controlled than with towed rafts, which swing freely at the end of a 50-m towline behind the pilot boat. When extracting by floating, two or more logs are usually tied together and, where possible, whole trees are topped and limbed, and then floated using manpower to a storage and rafting center. A motorboat is sometimes used to drag a line of logs attached end to end by chains, or by short cables. For heavy species or in shallow water, the yarding has to be done by cable, or, if the ground can support them, by high-flotation tracked skidders.

9.2.1
Mangrove and Tidal Forests

When located in tidal areas, the tree roots emerge more or less completely at low tide, and access roads have to be opened up by axe and bushknife. Numerous stilt roots, sometimes 1 or 2 m in height, cover the ground and leave no passage between the trees. There are also densely growing air roots, called pneumatophores, just high enough to emerge with their round tops out of the usual high water level, but not strong enough to support man or animal.

Often extraction of wood and wood products is still done by hand, with the use of small paddle boats for transportation. Mechanization of transport by water and by land depends on the type of existing waterways, whether shallow or deep water, whether wide or narrow, the tidal variations of the water level, and the speed of the incoming or outgoing tides. These determine the size and draft of the boats and barges, and also the mechanical equipment they can carry. Such equipment can be mounted permanently on decked barges or scows, and progressively shifted to new logging sites; or it can be rigged up on the shore, on platforms supported by log mats above the high water level.

Trees are cut at low tide, and the timber is pushed along with poles at high tide to deeper water, and then bound into rafts. For timber species which do not float when green, it is often possible to obtain temporary floatability by girdling the trees a few weeks, or a few months, before felling. If labor is scarce, the yarding production can be increased by the use of small motorboats with flat bottoms, for pushing or pulling the logs.

For greater production of nonfloating timber, a motor-driven yarder, mounted on a decked barge, is generally used for yarding and loading. The sheave blocks for the main and haulback cables are fixed to a strong tree near the barge, well-anchored to hold its position during both the rise and the fall of the tide. The sheave block, to return the cable, is fastened to a tall tree in the forest about 250–300 m away. Depending on the yarder capacity and on the timber sizes, logs or whole trees are skidded to the barges and loaded.

9.2.2
Swamps

The small fluctuation in water level in swamps simplifies the choice of working methods and of the required mechanical equipment. Swamps are generally the undrainable low parts of seasonally inundated areas. They do not contain the same wood species as mangrove forests, but contain mostly slow growing durable, heavier hardwoods.

9.2.3
Seasonally Inundated Forests

Working methods and mechanical equipment in seasonally inundated forests depend first on whether the logged timber is a floating species or not, and second on the total volume of the planned output. Floating timbers are cut during the dry season and are hauled out to deep water by animal, manpower, or mechanical transportation. On the riverbank or by lakeshores, the timber is made up into rafts and floated downstream to a shipping port or to a sawmill. In small creeks and rivers the high floodwater may last for only a very short period and, if all timber cannot be put into the water during this time, it has usually to wait on the shore for a whole year for the next flood. Nonfloating timber has to be hauled over land to shipping ports, and on the usual low-quality hauling roads this is best carried out during the dry season unless forest railroads are available.

9.2.4
Driving in Creeks and Rivers

Log driving down creeks and rivers is used for transport when creeks and rivers are too narrow, too fast, or otherwise unfit for the passage of rafts. The logs are floated down the stream to larger rivers or lakes where rafts can be built for further transportation.

Driving large tropical timber in creeks where the water level may rise or fall at the speed of 1 m/h requires good organization in order to profit from the short duration of suitable water levels. The system only works well if the necessary crews are available in time to roll the logs into the water, and if the high water lasts long enough to get all the timber away. To avoid log jams, the logs are often guided by men with long poles.

Depending on the frequency and the intensity of the rains, and on the speed the water is running, sometimes all logs cannot be floated in time or they have been grounded during the drive and have to wait 6–8 months for the next wet season before there is another chance for driving. Some species may not lose much of their quality during the first season, and some durable timber species may not suffer at all if not exposed too long to the sun. However, if timber cannot be floated during the second wet season and has to wait a whole year more, it can become a complete loss.

9.2.5
Rafting in Rivers and Lakes

Rafts can be bound with vegetable fiber, such as lianas and rattan, or with cable or chains. Rafting long logs in booms is often unsuitable for tropical logs because of the need to negotiate frequent bends in the waterways. A wire, pulled through the rings of spikes which have been driven into the end of each log, is sufficient to hold the logs together and at the same time gives the necessary flexibility for going around river bends.

Longer logs, rafted parallel to the stream direction, are attached to pole crossties, having the same length as the width of the raft. If sections of the same log length are formed, several sections are often tied together to form a long raft. Logs of different length, on the other hand, are attached to the necessary number of riders, and are usually bound into a single long solid raft, with only one layer of logs. If the buoyancy of the floating timber permits it, sawn timber can be transported as a top load. Nonfloatable timber can also form top loads or be bound between floaters.

Along large slow rivers, the necessity for building more rigid rafts for pushing, which would increase transportation costs, has prevented a general change from the towing system.

Many forests in the Far East from India to Indonesia include big pure bamboo reserves, which are harvested every 3–8 years. Small rafts of bamboo are made up of 50 poles, 8–10 m in length, attached together in three layers at the thicker leading end and spreading out fanwise at the rear end.

Several such units may be assembled together, laid one on top of the other, without any binding. The rafts are floated down the river with the current, guided and pushed by a few men equipped with long poles. They advance only a few kilometers per day and sometimes get stranded at low water. They arrive at the mills at irregular intervals and the mills are forced to keep big stocks on shore to ensure continuous production. Bamboo keeps its floatability as long as its cellular structure is not destroyed, and losses during the rafting are limited to bamboo sticks smashed or broken in collisions. Stranded rafts are disengaged from the rest of the floating ones and left to wait for high water.

9.3
Railroads

At one time, railways were a principal method of moving heavy timber from the forest to a river landing or to the sawmill. Temporary narrow-gauge railroads were common in the tropics. Round wooden ties were cut from the adjacent forest. Ties were not treated and were not flattened on the bottom and sometimes not on top and were laid directly on the ground with larger ties dug into the ground. Over swampy ground, longer ties were used to increase support. Rails were spiked directly to the ties. Locomotives from 7–22 t are used to pull low flat cars of 6–10-t capacity.

With the development of modern road construction and maintenance techniques and more powerful and reliable trucks, transport by railway has declined. In areas where well-planned and well-built railroads are already in existence they can often compete with trucks because their construction costs were written off long ago. In other areas, shortage of manpower for railroad maintenance has caused their abandonment.

Railways are most likely to compete with roads when:

- Forests are managed on a sustained-yield basis so that an adequate railway traffic level can be maintained.
- There are swamps to cross where rail construction is less expensive than either road construction across the swamp or long roads around the swamp.
- The transport distance is long enough to justify rehandling of the wood.
- There is the possibility to include cargo from other sources.

Culture, infrastructure, labor availability, and cost structures vary widely across countries with tropical forests. The appropriateness of a harvesting technology depends on a large number of economic, social, and political factors. These include:

1. Availability of unskilled labor and skilled labor
2. Cost of unskilled labor and skilled labor
3. Culture (work hours, work season)
4. Physical ability to do the work (log size, transport distance, slope, heat, humidity, snakes)
5. Cost per cubic meter produced by manual, animal, motor–manual, and mechanized systems
6. Labor laws (ability to terminate, employee benefits)
7. Social infrastructure (housing, schools, hospitals)
8. Possibility of labor unrest
9. Availability of capital
10. Equipment and parts import limits
11. Availability of parts, maintenance facilities
12. Prospective changes in labor laws
13. Prospective changes in equipment or fuel costs (subsidies)
14. Availability of training for the work force
15. Ability to retain the work force

In natural forests, the physical ability to fell large trees and skid large logs has led to widespread use of the chainsaw and either rubber-tired or tracked skidders and some form of machine-assisted truck loading. In planted forests there is a much wider choice (Table 10.1).

The introduction of machine-intensive technologies is most feasible where labor costs are high, daily and annual production requirements are high, material freshness is important, equipment maintenance and repair are reliable, fuel and spare parts are available on a continuous basis, and workers and foremen

Table 10.1. Options for stump to truck operations in planted forests

Fell	Buck	Skid	Process	Load
Bow saw	Bow saw	Animal		Manual
Bow saw	Bow saw	Farm tractor		Manual
Chainsaw	Chainsaw	Farm tractor		Hydraulic
Chainsaw	Chainsaw	Skidder		Hydraulic
Chainsaw	Chainsaw	Shovel		Hydraulic
Feller-buncher		Shovel	Processor	Hydraulic
Chainsaw		Skidder	Processor	Hydraulic
Feller-buncher		Skidder	Processor	Hydraulic
Feller-buncher		Clambunk	Processor	Hydraulic
Chainsaw	Chainsaw	Forwarder		Hydraulic
Harvester	Harvester	Forwarder		Hydraulic

A feller-buncher has a hydraulically powered chainsaw, a continuous rotating disc, or shears mounted on a rubber-tired or tracked carrier. A harvester is a hydraulically powered felling and processing head on a rubber-tired carrier, a tracked carrier, or an excavator chassis. A farm tractor can be equipped with a grapple or chokers, or can pull a trailer. A skidder can be equipped with a grapple or chokers, can be rubber-tired or tracked. A clambunk is a rubber-tired or tracked undercarriage with an inverted grapple for skidding tree lengths. A forwarder can be four-wheeled, six-wheeled, or eight-wheeled. A shovel is a modified tracked excavator chassis with a special loading boom. A processor is hydraulically powered delimbing and crosscutting machine mounted on a tracked or rubber-tired carrier. Hydraulic loaders can be tractor-mounted, truck-mounted, or on an excavator chassis. For long logs, rubber-tired front-end loaders might be used.

are skilled in the operation of expensive equipment. Generally mechanization is most effective when supported by a strong infrastructure of roads, communications, mechanical services, and managerial oversight.

The choice of the appropriate level of mechanization is challenging, regardless of whether the enterprise is private or publicly owned, and the target is a moving one. As the standard of living changes, the appropriate level of mechanization also changes. How quickly a society can incorporate changes in mechanization varies. Some enterprises choose to mechanize early to avoid potential disruptions from changing technology later.

As an example, assume a technology choice for a medium-sized plantation producing 500,000 m^3/year of eucalyptus pulpwood, 0.2 m^3 per tree, has two choices: (1) mechanized felling and bucking or (2) manual chainsaw felling and bucking under the following assumptions. A mechanized harvester produces 24 m^3/h and costs $80 per hour. A chainsaw costs $2.5 per hour to own and operate. Felling and bucking production with a chainsaw is 2 m^3/h. What is the breakeven cost of chainsaw labor including supervision and social costs? If the cost of a cutter including supervision and social benefits is less than $5.00 per hour, then manual cutting with a chainsaw yields a lower cost per cubic meter. If the cost is greater than $5.00, then mechanized cutting provides a lower cost per cubic meter (Fig. 10.1). But, there are other considerations. Management will need to provide supervision for 12 times as many men. Manual cutters cannot work at night and manual cutting will have higher accident rates than mechanized felling. Work stoppages with manual cutting are more difficult to manage. Manual cutting also does not permit short-term boosts in production that mechanized felling can provide by going to two, two and a half, or even three shifts per day. On the other hand, manual cutting provides rural employment, increases community participation, and may create economic development pathways. In some cases, the cost of public services may be reduced if the alternative to employment is unemployment and public subsidies. In our example, if the cost of a chainsaw operator is $3.00 per hour and the enterprise choses to mechanize to avoid potential labor problems, the opportunity cost of that decision is approximately $1 per cubic meter or $500,000 per year for 500,000 m³. The rate of mechanization in temperate countries has been driven by labor availability, labor cost, and equipment technology, most recently by advancements in hydrostatic systems. Planted forests in Scandinavia have tended toward harvester and forwarder systems for both thinnings and final harvests. In eastern North America, feller-bunchers and grapple skidders or clambunk skidders, excavators with processing heads and forwarders, and harvesters and forwarders are used. In western North America where terrain is suitable, shovel logging, feller-bunchers with grapple skidders, and harvesters and forwarders are used; otherwise, skyline yarding often with a

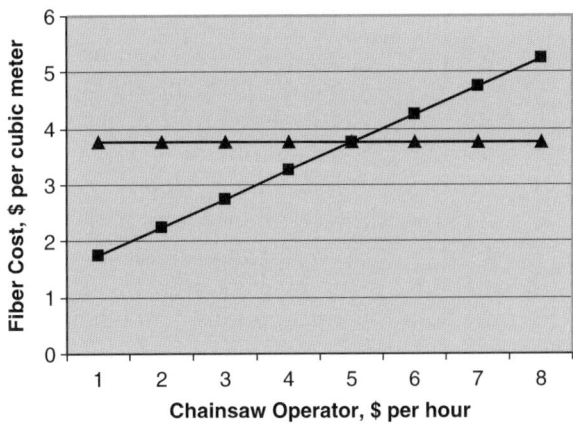

Fig. 10.1. Manual chainsaw cutting as a function of chainsaw operator labor cost compared with mechanized cutting at $3.75 per cubic meter. The breakeven chainsaw operator cost is $5.00 per hour

processor on the landing is used. Almost all loading is done by hydraulic loaders. Shovel logging is restricted to final harvests.

In the tropical and subtropical forests of South America, operations range from chainsaw felling and bucking with skidding by agricultural tractors to harvester and forwarder operations. Recent trends have been toward use of a processing head on an excavator chassis with transport to the roadside by forwarder. In tropical and subtropical Africa, operations range from bow-saw cutting and animal logging to chainsaws with agricultural tractor transport to the roadside. Recent concerns over availability of the future work force owing to HIV/AIDS has prompted interest in increased mechanization.

In tropical and subtropical Asia, operations range from chainsaw felling and bucking with skidding by rubber-tired skidders and yarding on steep areas with cable cranes.

This book has concentrated on describing harvesting systems and harvesting methods suitable for natural and planted forests of the tropics. It is published at a time of intense discussion concerning the importance and future of forests in tropical regions. The past 20 years has seen a significant expansion in tools designed to improve forest management, to ensure better wood recovery from harvesting operations, and to minimize forest damage arising from logging operations. These tools include criteria and indicators for sustainable forest management, certification standards for forest management, principles and requirements for compulsory forest management plans, reduced impact logging techniques, and codes of practice for forest management and forest harvesting. Particular attention has been focused on logging in natural tropical forests. Some countries have moved toward banning all harvests in natural forest owing to the inability to control logging impacts. While the merits of that action are not debated here, what does seem appropriate is to work toward "banning" poor logging practices. There is a substantial body of knowledge to support logging practices that minimize environmental impacts. It requires proper planning, efficient implementation, and a well-trained work force. The challenge ahead in natural forests is to establish the regulatory and educational environment which encourages the implementation of what is known and is increasingly being demonstrated.

Planted forests in the tropics present a different set of challenges. Planted forests serve a variety of purposes ranging from sources of industrial materials to energy to watershed protection. Where harvesting is involved, the choice of the appropriate level of harvesting technology is an important decision. The tremendous biophysical, social, political, and economic diversity among tropical countries requires different approaches to economic development. Labor-intensive technologies may be highly appropriate in one region, while mechanization may be more appropriate in another. Often it involves choosing the path of development which recognizes labor availability, training, education, and availability of capital. We have attempted to describe the various available technologies and their uses in this book.

References

Dawkins HC (1959) The volume increment of natural tropical high-forest and the limitations of its improvement. Emp For Rev 38(2):175–180

Dykstra D (1996) FAO model code of forest harvesting practice. FAO. Rome.

Hammond D (ed) (1995) Forestry handbook. New Zealand Institute of Forestry, Christchurch

Holmes T, Boltz F, Blate G, Zweede J, Perreira R, Barreto P, Boltz F, Bauch R (2000). Financial costs and benefits of reduced-impact logging relative to conventional logging in the eastern Amazon. Tropical Forest Foundation, Washington

FAO (1974) Logging and log transport in tropical high forest. A manual on production and costs. FAO, Rome

FAO (1976) Harvesting planted forests in developing countries. A manual on techniques, roads, production and costs. FOI: TF-INT 74 (SWE). FAO, Rome

FAO (1977) Planning forest roads and harvesting systems. FAO forestry paper 2. FAO, Rome

FAO (1999) Code of practice for forest harvesting in Asia-Pacific. RAP publication 1999/12. FAO and Asia-Pacific Forestry Commission. http://www.fao.org/docrep/004/ac142e/ac142e00.htm

FAO (2005) Regional code of practice for reduced-impact forest harvesting in tropical moist forests of west and central Africa. FAO, Rome

FAO/ILO (1980) Chainsaws in tropical forests. FAO/ILO, Rome

Kantola M, Harstela K (1988) Handbook on appropriate technology for forestry operations in developing countries. Part 2. Wood transport and road construction. Forestry Training Programme publication 19. National Board of Vocational Education of the Government of Finland, Helsinki

Further Reading

Brown C, Sessions J (1999). Variable tire pressures for tropical forests? A synthesis of concepts and applications. J Trop For Sci 11(2):380–400

Cermak FI, Lloyd AH (1963) Timber transportation in the tropics. FAO, Rome

Chandra R (1975) Production and cost of logging and transport of bamboo. FAO/SWE/TF 157. FAO, Rome

Dykstra D (2003) RILSIM 2.0 user's guide. Software for financial analysis of reduced-impact logging systems. USDA Forest Service

FAO (1981) Cable logging systems. FAO forestry paper 24. FAO, Rome

FAO (1982) Basic technology in forest operations. FAO forestry paper 36. FAO, Rome

FAO (1984) Self-loading winch trucks – based on the work of JL Wilson. JG Groome and Associates, Rome

FAO (1987) Appropriate wood harvesting in plantation forests. Training materials from the FAO/Finland training course on appropriate wood harvesting operations, Mutare, Zimbabwe, 9–26 June 1986. FAO forestry paper 78. FAO, Rome

Folkema MP, Hermelin J, Saunders J (1977) Handbook for logging with farm tractor-mounted winches. Forest Engineering Research Institute of Canada, handbook 2. Forest Extension Service, New Brunswick

Forestry Training Programme (1988) National Board of Vocational Education. Appropriate forest operations. Proceedings of the FAO/Finland training course, Philippines, 23 November–11 December 1987. FAO, Rome

Garland JJ (1983) Designated skid trails minimize soil compaction. Extension circular 1110. Oregon State University, Corvallis

Hakkila P, Malinovski J, Siren M (1992) Feasibility of logging mechanization in Brazilian forest plantations. Finnish Forest Research Institute, Helsinki

Kantola M, Virtanen P (1986) Handbook on appropriate technology for forestry operations in developing countries. Part 1. Tree felling and conversion clearing of forest plantations. Forestry Training Programme publication 16. National Board of Vocational Education of the Government of Finland, Helsinki

Lysons H, Mann C (1967). Skyline tension and deflection handbook. PNW-39. USDA Department of Agriculture, Forest Service

Oregon State University LoggerPC 4.0. Forest Engineering Department, Oregon State University, Corvallis

Sessions J, Heinrich R (1993) Harvesting. In: Pancel L (ed) Tropical forestry handbook. Springer, Berlin Heidelberg New York, pp 1325–1424

Index